Acknowledgements

The National Task Group wishes to acknowledge the comment ('My thinker's not working') by Ray, a 60ish man residing in a Kansas group home, on how he perceived the effects of his early dementia.

Primary support for the National Task Group was provided by the American Academy of Developmental Medicine and Dentistry, the Rehabilitation Research and Training Center on Aging with Developmental Disabilities-Lifespan Health and Function at the University of Illinois at Chicago, and the American Association on Intellectual and Developmental Disabilities. The support provided by the American Association of University Centers on Disabilities is also acknowledged.

This document was produced under grant number H133B080009 awarded by the U.S. Department of Education's National Institute on Disability and Rehabilitation Research to the Rehabilitation Research and Training Center on Aging with Developmental Disabilities-Lifespan Health and Function at the University of Illinois at Chicago. The contents of this article do not necessarily represent the policy of the U.S. Department of Education, and should not be assumed as being endorsed by the U.S. Federal Government.

I0392583

Citation

National Task Group on Intellectual Disabilities and Dementia Practice. (2012). *'My Thinker's Not Working': A National Strategy for Enabling Adults with Intellectual Disabilities Affected by Dementia to Remain in Their Community and Receive Quality Supports.*
www.aadmd.org/ntg/thinker ● www.rrtcadd.org/ ● www.aaidd.org

v. January 2012

National Task Group on Intellectual Disabilities and Dementia Practice
Co-Chairs
Matthew P. Janicki, Ph.D. and Seth M. Keller, M.D.

TG Steering Committee members contributing to this report

Kathy Bishop, ...D., Melissa DiSipio, MSA, Lucy Esralew, Ph.D., Lawrence T. Force, Ph.D., Mary Hogan, MAT, Nancy Jokinen, Ph.D., Ronald Lucchino, Ph.D., Philip McCallion, Ph.D., Julie A. Moran, D.O., Leone Murphy, MSN, Linda Nelson, Ph.D., Dawna T. Mughal, Ph.D., Nabih Ramadan, M.D., Kathy Service, Ph.D., Baldev K. Singh, M.D.

Executive Summary

'My Thinker's Not Working': A National Strategy for Enabling Adults with Intellectual Disabilities Affected by Dementia to Remain in Their Community and Receive Quality Supports, prepared by the National Task Group on Intellectual Disabilities and Dementia Practices, provides a summary of the challenges facing the nation as we observe an increasing rate of dementia found in older people with intellectual disabilities. The Report offers recommendations for the various stakeholders in the field of intellectual disabilities and anticipates that its findings and recommendations will be considered and integrated into the annual reports and plans developed by the federal Advisory Council on Alzheimer's Research, Care, and Services.

The National Task Group on Intellectual Disabilities and Dementia Practices first and foremost recognizes that the number of older adults with an intellectual disability affected by dementia is growing and this growth is posing a significant challenge to families and friends, provider agencies, and federal and state agencies concerned with supports and services to people with an intellectual disability. It also recognizes that although the research community is making significant strides in better understanding the causal and evolutionary factors leading to the onset of Alzheimer's disease and other dementing illnesses and is also making significant progress in identifying means for the early detection of the disease – all of which will benefit subsequent generations – the social care system still remains challenged with the 'here and now' of offering the best and most efficacious means of identification, daily supports, and long-term care.

The key findings of the National Task Group include:

● Most adults with an intellectual disability live in community settings, either independently or with support from families, friends and service providers; with advanced age, they may experience age-related conditions and diseases, including dementia.

● Epidemiological research has not arrived at reliable population counts of adults with an intellectual disability affected by mild cognitive impairment and dementia and more effort is needed to create a more reliable estimate of this population.

● Dementia has a devastating impact on adults with an intellectual disability as well as on their families, friends, housemates, and service provider staff who often provide key long-term support and care.

● Community services' providers are facing a 'graying' of their service population, many of whom are affected by cognitive decline and dementia, and are challenged to provide the most effective and financially viable daily supports and long-term care.

● Primary care and supports for adults with an intellectual disability affected by dementia can be primarily provided within the community and appropriate services can preclude institutionalization.

● Providers are beginning to adapt small group homes for specialized community care and supports for persons with an intellectual disability affected by dementia.

● Professional staff are often ill-equipped to help identify and support interventions that may be the most efficacious for adults with an intellectual disability affected by dementia.

• There is a lack of background knowledge and training in late life problems of adults with an intellectual disability among primary care health providers (including physicians, physician assistants, and nurses) in community practice.

• Specialized assessment and diagnostic resources are needed to help more effectively identify adults with an intellectual disability and dementia.

• A common screening instrument would be useful for the cognitive impairment review that is part of the Affordable Care Act's annual wellness visit.

• Creating a national program of trainings using workshops, webinars, and other teaching methods, would advance the knowledge and skills among workers and clinicians working with adults with an intellectual disability affected by dementia.

• Creating a national information and education program for adults with an intellectual disability and family members would improve their understanding of dementia and potentially lead to earlier identification and acquisition of timely supportive services.

• Access to appropriate professionals and supportive services outside major urban settings needs to be improved; technology may play an important role in achieving this goal.

• State and local developmental disabilities' authorities could more constructively forecast and budget for supporting in-community care of adults with an intellectual disability affected by dementia.

The Report concludes with a series of recommendations that comprise a **National Action Plan** (see page iv) for more effectively addressing needs and helping adults with an intellectual disability affected by dementia. Summarized below are some of the main areas that are covered by the recommendations.

Dementia often hits harder. Alzheimer's disease and other dementias generally affect adults with lifelong intellectual disabilities in similar ways as they do other people, but sometimes have a more profound impact due to particular risk factors – including genetics, neurological injury, and deprivation. While such illnesses generally follow a typical course in terms of impact and duration, some adults are profoundly and aggressively affected. Yet all need the typical types of supports and services usually associated with dementia-capable care. *The National Task Group believes that adults with an intellectual disability require the same early and periodic diagnostic services, community education, and community-based supports for themselves, their caregivers, and the organizations working with them, as do other adults affected by dementia.*

Lifelong caregiving may create 'double jeopardy'. Many families are the primary lifetime caregivers for adults with an intellectual disability and when Alzheimer's disease and these dementias occur, they are particularly affected and need considerable supports. These families not only include parents, but also siblings and other relatives. Many such families are at a loss for providing extensive care at home once dementia becomes pronounced and care demands may overwhelm them. *Thus, the National Task Group recommends that the nation's providers and federal and state aging and developmental disabilities authorities invest in increased home-based supports for caregivers who remain the primaries for support and care for adults affected by dementia.*

Providers are being challenged. Many intellectual disabilities' provider organizations that are the primary resources for residential and day supports are vexed by the increasing numbers of adults with an intellectual disability in their services showing signs of early decline and dementia with potentially more demanding care needs. In many cases, staff may be unfamiliar with the signs and symptoms of mild cognitive impairment (MCI) or dementia and may misrepresent or ignore these changes, when early identification and intervention could prove beneficial. ***Thus, the National Task Group recommends that the nation's providers and federal and state aging and developmental disabilities authorities invest in increased education and training of personnel with respect to Alzheimer's disease and other dementias and invest in promoting best practices in models of community care of adults with an intellectual disability affected by dementia.***

Early identification is crucial. As it is important to recognize signs of dementia-related cognitive decline early on, the National Task Group has identified a potentially adaptable instrument, applicable particularly to adults with an intellectual disability, which can be utilized as a 'first-instance screen' and recommends adoption of such an instrument by providers and regulatory authorities to identify those adults at-risk due to early signs of mild cognitive impairment (MCI) or dementia. ***Thus, the National Task Group recommends that the nation's providers and health authorities undertake a program of early identification – beginning at age 50 for adults with an intellectual disability and at age 40 for adults with Down syndrome and others at early risk – using a standard screening instrument.***

Commitment to living in the community. Research has shown that community-based models of care for adults with an intellectual disability and dementia including community-based options, such as support for living at home or in small group homes, are viable and gaining preference for all individuals affected by Alzheimer's disease and other dementias. The institutionalization of adults with an intellectual disability and dementia is anathema to the field's core beliefs and commitments to care practices; institutionalization (via use of long-term care facilities) can have an adverse effect on lifespan and quality of life. ***Thus, the National Task Group recommends that the use of such community-care options be expanded and an investment be made in developing more small community-based specialized 'dementia capable' group homes.***

Education is what's missing. Information at all levels is needed to enhance the capabilities of staff, clinicians, community providers and administrators. Training of various sorts is necessary to raise awareness of dementia and how it affects adults with an intellectual disability. The National Task Group recognizes the need for more information related to age-associated cognitive decline and neuropathologies (such as dementia), particularly how they apply to people with an intellectual disability and impact their families, friends, advocates and caregivers. ***The National Task Group recommends the institution of a national effort on training and education to prepare the workforce and eliminate disparities in dementia services provision for adults with an intellectual disability.***

A final word. Dementia has a devastating impact on all people – including people with an intellectual disability and their friends, families and the staff who may be involved with them as advocates and caregivers. ***The National Task Group believes that the federal Advisory Council on Alzheimer's Research, Care, and Services should include concerns and considerations for people with lifelong intellectual disabilities in any and all documents, plans, and recommendations to Congress that are part of the work of the Council through to 2025.*** To this end, the National Task Group stands ready to assist and contribute to such efforts.

What follows is a matrix listing the National Task Group's recommendations as to what should be undertaken and which organization or group could be involved.

National Dementia and Intellectual Disabilities Action Plan

Goal A: To better understand dementia and how it affects adults with an intellectual disability and their caregivers

Number	Recommendation	Who could do it?
#1	Conduct nationwide epidemiologic studies or surveys of adults with intellectual disabilities that establish the prevalence and incidence of mild cognitive impairment and dementia.	Federal agencies and institutes (Administration on Developmental Disabilities, Administration on Aging, National Institute on Disability and Rehabilitation Research)
#2	Conduct studies to identify and scientifically establish the risk factors associated with the occurrence of dementia among adults with an intellectual disability.	Universities' academic and research centers
#9	Conduct studies on the impact of aging of family caregivers on the support and care of adults with intellectual disabilities residing in at-home settings.	Universities' academic and research centers
#11	Conduct nationwide medico-economic studies on the financial impact of dementia among people with intellectual disabilities in various service provision settings.	Universities' academic and research centers

Goal B: To institute effective screening and assessment of adults with an intellectual disability at-risk, or showing the early effects of, dementia

Number	Recommendation	Who could do it?
#3	Develop guidelines and instructional packages for use by families and caregivers in periodically screening for signs and symptoms of dementia.	American Academy of Developmental Medicine and Dentistry
#4	Encourage provider agencies in the United States to implement screenings of their older-age clientele with an intellectual disability who are at-risk of or affected by dementia.	State developmental disabilities planning councils, State developmental disabilities authorities
#5	Examine the utility of adopting an instrument such as an adapted Dementia Screening Questionnaire for Individuals with Intellectual Disabilities for use annually in preparation for the annual wellness visit.	Universities, Providers, American Academy of Developmental Medicine and Dentistry
#6	Conduct an evaluation of a workable scoring scheme for the Dementia Screening Questionnaire for Individuals with Intellectual Disabilities that would help identify individuals in decline.	Universities' academic and research centers
#8	Promote the exchange of information among clinicians regarding technical aspects of existing assessment and diagnostic instruments for confirming presence of dementia in persons with an intellectual disability.	American Academy of Developmental Medicine and Dentistry, American Association on Intellectual and Developmental Disabilities, Developmental Disabilities Nurses Association

Goal C: To promote health and function among adults with an intellectual disability

Number	Recommendation	Who could do it?
#15	Develop and disseminate a set of nutritional and dietary guidelines appropriate for persons with an intellectual disability affected by dementia.	American Academy of Developmental Medicine and Dentistry
#16	Develop and disseminate health practice guidelines to aid primary care physicians and related health practitioners address assessment and follow-up treatment of adults with an intellectual disability presenting with symptoms of dementia.	American Academy of Developmental Medicine and Dentistry, Developmental Disabilities Nurses Association
#17	Conduct studies on the nature and extent of health compromises, conditions, and diseases found among adults with an intellectual disability and affected by dementia.	Universities' academic and research centers

Goal D: To produce appropriate community and social supports and care for adults with an intellectual disability affected by dementia

Number	Recommendation	Who could do it?
#10	Enhance family support services to include efforts to help caregivers to identify and receive assistance for aiding adults with an intellectual disability affected by dementia.	State developmental disabilities authorities, State units on aging, Area agencies on aging, The Arc, National Down Syndrome Society
#12	Plan for and develop more specialized group homes for dementia care as well as develop support capacities for helping adults affected by dementia still living on their own or with their family.	State developmental disabilities authorities
#13	Plan and develop community-based dementia-capable supports to address the needs of those persons at-risk or affected by dementia.	State developmental disabilities authorities
#14	Develop and disseminate social care practice guidelines to community agencies and professionals that address assessment, service development and life planning for adults with an intellectual disability presenting with symptoms of dementia.	American Association on Intellectual and Developmental Disabilities

Goal E: To produce a capable workforce and produce education and training materials

Number	Recommendation	Who could do it?
#7	Establish undergraduate, graduate, and continuing education programs, using various modalities, to enhance the diagnostic skills of community practitioners.	American Academy of Developmental Medicine and Dentistry, American Association on Intellectual and Developmental Disabilities, Council of Deans of Medical Schools and Allied Health Colleges
#18	Develop a universal curriculum, applicable nationwide, on dementia and an intellectual disability geared toward direct care staff, families, and other primary workers.	Administration on Developmental Disabilities, Universities, Developmental Disabilities Nurses Association
#19	Organize and deliver a national program of training using workshops and webinars, as well as other means, for staff and families.	American Academy of Developmental Medicine and Dentistry, American Association on Intellectual and Developmental Disabilities, Developmental Disabilities Nurses Association, Universities' academic and research centers
#20	Develop and produce an education and information package for adults with an intellectual disability to help them better understand dementia.	American Academy of Developmental Medicine and Dentistry, Developmental Disabilities Nurses Association, Universities' academic and research centers

'My Thinker's Not Working'

This document is dedicated to all of the men and women with intellectual disabilities who were, are, or may possibly be faced with the challenge of Alzheimer's disease or related dementias

Collaborating Organizations

Alzheimer's Association
Alzheimer's Disease International
American Academy of Developmental Medicine and Dentistry
American Occupational Therapy Association
American Association on Intellectual and Developmental Disabilities
Association of University Centers on Disabilities
Developmental Disabilities Nurses Association
Down Syndrome International
IASSID Quality of Life Special Interest Research Group
Mt. St. Mary College's Center on Aging and Policy
National Association of State Directors of Developmental Disabilities Services
National Down Syndrome Society
The Arc
University at Albany's Center on Excellence in Aging
University of Illinois at Chicago's RRTC on Aging with Developmental Disabilities–
Lifespan Health and Function
University of Northern British Columbia's School of Social Work
University of Rochester's Program on Aging and Developmental Disabilities

Table of Contents

Foreword

From the 1940s, when the first published reports began to appear that some adults with an intellectual disability were affected in later age by debilitating decline, there has been growing interest in what is the nature of this decline, who does it effect, and what may be its causes. Successive work identified Alzheimer's disease as the main cause and many studies began to examine how it affected adults with Down syndrome, which was the condition most commonly affected. Over time and worldwide, a variety of researchers successively contributed more information to these findings. A first effort to bring together key researchers to review and synthesize the issues related to the occurrence of dementia among adults with intellectual and developmental disabilities was a NICHD funded meeting held in 1994 in Minneapolis, Minnesota, in conjunction with the 4th International Conference on Alzheimer's Disease and Related Disorders.

This international meeting was the first time many of world's researchers, clinicians, and advocates concerned with this topic met to examine the state of the science and propose new directions for research and practice. From this meeting, among other products, emerged the 'AAMR/IASSID Practice Guidelines' which provided a structure for stage-related care management of Alzheimer's disease among adults with intellectual disabilities, and offered suggestions for the training and education of caregivers, peers, clinicians, and program staff. These guidelines also served to suggest public policies that reflected a commitment to aggressive care of people with Alzheimer's disease and intellectual disability, and avoidance of institutionalization solely because of a diagnosis of dementia.

The National Task Group wishes to acknowledge the contribution of the 1994 meeting and its varied products, as well as all of the other efforts undertaken in the United States and elsewhere that have helped bring us, with respect to knowledge and policies, to where we are today. Specifically, the National Task Group wishes to acknowledge the efforts of Commissioner Sharon Lewis of the Administration on Developmental Disabilities (ADD) for convening numerous ADD Listening Forums in 2010 during which the need for a national effort on dementia and intellectual disabilities was explored, and the efforts of family advocates, such as Mary Hogan, Leone Murphy, and others, who have brought this issue to the forefront of the nation's conscience. It also acknowledges the leadership of Dr. Seth Keller, President of the American Academy of Developmental Medicine and Dentistry, and Dr. Matthew Janicki, of the Rehabilitation Research and Training Center on Aging with Developmental Disabilities at the University of Illinois at Chicago, who organized, facilitated, and shepherded the work of the National Task Group.

It is the hope of the National Task Group that its efforts will contribute significantly to the national conversation on how to best address the effects of this insidious disease among one notable group of Americans.

Matthew P. Janicki, Ph.D., & Seth M. Keller, MD, Co-Chairs

National Strategy for Enabling Adults with Intellectual Disabilities Affected by Dementia to Remain in Their Community and Receive Quality Supports

---

Bill's sister provided us with his story, which typifies a common situation

"Bill, who had Down syndrome, continued to live at home with our mother after the rest of us had grown up and moved away. When he was about 30, Bill moved to a group home and had a job at a local workshop. After our mother's death, my sisters and I became more involved with planning for his future. We worked with other families and the group home's staff to plan long term for 'aging in place'. When Bill reached his mid-forties he first began to have cognitive and physical problems and a medical diagnosis of Alzheimer's disease was made. As these problems increased, the agency made adaptations to Bill's work and his home. However, he experienced more rapid decline as well as behavioral changes by his late-forties. Initially, additional staff were added to his group home to meet his needs, but his condition deteriorated further and concerns about his safety increased. We and the agency wanted to avoid a referral to a nursing facility, so together we planned for a transition to a group home serving multiply-involved adults. Even though we worked with the staff to help Bill make the transition, it may have not been the right setting for him – as the staff were unfamiliar with dementia. Unfortunately, he deteriorated and at age 49 died of complications related to Alzheimer's disease – some eight months after the move."

1.0 Charge and Focus

It is expected that efforts undertaken by the US Department of Health and Human Services with respect to the requirements of the National Alzheimer's Project Act, Public Law 111-375 (42 U.S.C. 11225) - 'NAPA' - enacted by Congress in 2011, will lead to the development of a coherent and coordinated national strategy on dealing with Alzheimer's disease in the United States. To complement this federal initiative and to address the myriad requests for more specific information and practice models for providing quality care for people with an intellectual disability affected by dementia, the American Association on Intellectual and Developmental Disabilities (AAIDD), the American Academy of Developmental Medicine and Dentistry (AADMD), along with the Rehabilitation Research and Training Center on Aging with Developmental Disabilities-Lifespan Health and Function at the University of Illinois at Chicago, created the National Task Group on Intellectual Disabilities and Dementia Practices (NTG).

The National Task Group, made up of some 100 administrators, academics, providers, clinicians, family members, and advocates (see 9.2), coalesced with a number of national disability and family based organizations, federal agencies, and provider representatives to work on a broad basis within three working groups. Besides working in the groups, the members of the National Task Group met in two plenary sessions, one in St. Paul, Minnesota in June 2011 and another in Arlington, Virginia in November 2011.

The mission of the National Task Group was to develop and disseminate an action plan that would both contribute to the NAPA effort and ensure that the concerns and needs of people with an intellectual disability and their families, when affected by dementia, were considered as part of such a national strategy. The intent would be that adults with an intellectual disability would receive the same access to services developed and provided by the States as anyone else affected by dementia.

When formed, the immediate goal of the National Task Group was to identify, define and promote the best technologic and clinical practices used by agencies in delivering supports and services to adults with an intellectual disability affected by various dementias. Equally, the National Task Group was asked to identify a workable screening instrument that would help identify and substantiate suspicions of dementia-related decline, suggest and promote the development of a set of practice guidelines for post-determination health care and supports, and recommend models of community-based support and long term care of persons with an intellectual disability affected by dementia.

It is hoped that this document will add to the national conversation on how the United States can address the impact dementia will have on its older population – and in particular those adults with lifelong intellectual disabilities. As its contribution, this document discusses and proposes a plan of action for enabling adults with an intellectual disability affected by dementia and/or mild cognitive impairment and their caregivers to receive supports appropriate for their needs and function. It is framed within the goal of promoting community inclusion, optimal independence, and maintaining quality of life to the highest degree possible.

2.0 The Population

2.1. Definition of Intellectual Disabilities

Persons who are the focus of this report are those adults having an intellectual disability who are affected by dementia. In this report, the National Task Group has adopted a practical definition[1] that characterizes adults with an intellectual disability (formally termed 'mental retardation') affected by dementia as those who:

- have intellectual limitations that significantly limit the person's ability to successfully participate in normal day-to-day activities such as self-care, communication, work, or going to school, and
- developed the intellectual limitation during the 'developmental period' (before approximately age 22), and
- the limitation is anticipated to result in long term adaptive or functional support needs, and/or
- are eligible for State or Federal public support programs because they have been diagnosed as having an intellectual disability; and
- are affected by dementia, and meet the criteria of having been diagnosed with possible, probable, or definitive dementia, or have mild cognitive impairment, as defined by the World Health Organization's International Classification of Diseases or meet the diagnostic criteria of the American Psychiatric Association's Diagnostic and Statistical Manual.

There are many causes of intellectual disabilities, some genetic, some hereditary, and some social or environmental. Among the genetic causes, Down syndrome is the one most commonly associated with dementia as adults with Down syndrome are at high risk of Alzheimer's disease and generally manifest early onset of dementia.

Dementia is a term that characterizes the progressive loss of brain function that occurs with certain neuropathological diseases or trauma and is often associated with aging. It is marked by memory disorders, personality and behavioral changes, and impaired reasoning. Dementia is not a disease itself, but rather a group of symptoms that are caused by various degenerative brain diseases or conditions, such as Alzheimer's disease, stroke, or other brain trauma. Alzheimer's disease is the most prevalent cause, associated with some two-thirds of the instances of dementia. There are different types of dementias, among them Alzheimer's dementia, vascular dementia, fronto-temporal dementia, Lewy body dementia are the most prevalent. Dementia related to a brain disease, such as Alzheimer's, generally has a progressive nature so that over time the individual affected continually loses more cognitive and functional skills and eventually is totally unable to function independently. The result is debilitation and death.

2.2 Estimates of Impact

The Alzheimer's Association estimates that currently some 5.2 million Americans are affected by dementia, many of whom have Alzheimer's disease. This disease is the sixth leading cause of death in the United States. Of this number some 200,000 affected adults are under the age of 65. In the United States, it is generally acknowledged that although persons with an intellectual disability are affected by dementia to the same degree as other adults in the general population [2,3], some may be affected earlier and at a greater rate. This includes adults with Down syndrome, most of who will be among those 200,000 adults affected under age 65. Generally, it is believed that about 6% of adults with an intellectual disability will be affected by some form of dementia after the age of 60 (with the percentage increasing with age). For adults with Down syndrome, studies show that at least 25% will be affected with dementia after age 40 and at least 50 to 70% will be affected with dementia after age 60. [2,4,5]

Currently in the United States, no national studies exist that have produced actual counts of persons with an intellectual disability diagnosed with dementia, although some national and state surveys have been undertaken to arrive at estimates. One national survey of state developmental disabilities agencies, conducted under the auspice of the National Association of State Directors of Developmental Disabilities Services, found that information on the extent of dementia from the state agencies' perspective is often rudimentary and sketchy. Thus, focal surveys conducted at the state or local level may provide a better indication of the numbers affected.

A comprehensive state survey in New York in the late 1990s found rates of dementia among adults with an intellectual disability comparable to that of the general population. [3] Although the report projected at the time that at minimum about 9,000 adults with an intellectual disability in the United States would be affected by dementia, it also noted that the number of such adults affected by dementia would most likely triple over the intervening years. As the survey captured information on known persons in service, it is reasonable to assume that there were additional persons affected by dementia and not in service who were not identified in the survey. Given the expected upward trend in prevalence as a result of population aging, it is likely that three times the number of adults, some 27,000 nationally, would be affected at this time. Also, doubling that number to potentially accommodate those persons not known to providers, on the margins of the intellectual disabilities definition, or adults with mild cognitive impairment (MCI) would provide a prevalence estimate of some 54,000 adults with an intellectual disability with mild cognitive impairment or dementia by the end of this decade.

Thus, it could be projected that there might be at least 54,000 adults with an intellectual disability and cognitive decline (dementia and MCI) in the United States. Further, estimates drawn from the *2010 State of State* report also suggest that there may be about 33,000 adults with *developmental* disabilities and dementia currently living at home with older family caregivers and perhaps twice as many living in out-of-family-home settings.[6] It may be difficult to parse the number of adults only with an intellectual disability from these broader data, but they provide a sense of the greater impact of dementia in the overall developmental disabilities population. While the number may seem modest with respect to the 5.2 million estimated by the Alzheimer's Association, these 54,000 would comprise a group with high dependencies and a high impact on caregivers, many of whom may have been providing lifelong care. They also represent adults whose needs pose a potential impact on national long term care resources as they would be Medicaid eligible. Recognizing that there may be a significant gap between the prevalence estimates and the actual number of individuals diagnosed – much like that in the general population – prudent use of any estimates is called for. The lack of reliable prevalence data points to the need for nationwide epidemiological studies that may produce more reliable estimates of the number of persons affected.

Such state or local surveys provide us with only estimates of the number of adults with an intellectual disability in the United States who may have dementia. What is known is that a significant percentage of this number will be adults with Down syndrome, who usually make up about 10% of most providers' service populations of adults age 40 and older. Accordingly, some agencies may experience a notable impact on service demand if they support a large proportion of older adults with Down syndrome. Thus, each organization will need to assess such impact on its client population by examining its demographics and deriving projections based on age, risk, and other salient factors. Current projections are that among adults with Down syndrome this means that at least two of every three older adults will eventually be affected by dementia. Among adults with an intellectual disability in general, this will mean that about one in eight older adults will be affected.

The Alzheimer's Association projects that the number of older persons affected by Alzheimer's disease will grow from about 5.2 million currently to about 7.7 million by 2030[7]. With increased life expectancy and the greater numbers of aging people due to the 'baby boom' effect, the number of older people will increase significantly over the next twenty years. Expectations are that this phenomenon will be mirrored among older adults with an intellectual disability. It is conceivable that any estimates of the number of persons with an intellectual disability currently affected will need to be at least doubled in projecting over the next twenty years.

Recommendation #1:
Conduct nationwide epidemiologic studies or surveys of adults with intellectual disabilities that establish the prevalence and incidence of mild cognitive impairment and dementia.

2.3 Risk Factors Specific to People with An intellectual disability

There are a number of specific risk factors that may lead to a higher prevalence of Alzheimer's disease or other conditions leading to dementia, including the presence of Down syndrome, significant head injuries, obesity, limited cognitive reserve, and poor cardiovascular health. The presence of Down syndrome is the most prominent risk factor among adults with an intellectual disability, increasing the prevalence to rates in excess of 60% of adults over age 60. No other cause of an intellectual or developmental disability is known to have the same risk for dementia as does having Down syndrome – although some other syndromes or conditions may also present with some elevated risk.

Studies have shown that adults with Down syndrome are at high risk for dementia, primarily caused by Alzheimer's disease.[8] The average age of onset (when we first suspect or begin to observe the changes in behavior associated with dementia) for adults with Down syndrome is about age 52; although some adults may show earlier onset (in their 40s and some in their 30s). As adults with Down syndrome have a lower life expectancy (their average age of death, after they reach age 40, is about 56 years), they also experience a compressed duration of symptoms. The illness duration (that is, life expectancy after recognized symptom onset) is about 5 to 8 years on average (compared to 7 to 20 years for adults with other types of intellectual disabilities or people in the overall population). Some adults with Down syndrome are subject to an aggressive form of Alzheimer's disease and will decline rapidly and die within about two years following 'onset'. In others, decline maybe witnessed as rapid because detection of early stage signs and symptoms is often late – so when dementia is finally identified progression appears accelerated.

Studies in the United States have also shown that adults with an intellectual disability with etiologies other than Down syndrome have a similar prevalence of

> *We were told of a situation of a woman who was affected by an aggressive form of Alzheimer's disease.*
>
> *"Ms. Janet Sudette was 49 years old when she died of complications related to Alzheimer's disease. Ms. Sudette had Down syndrome and lived for many years with her family before moving to a small group home. After a few years, she decided to move to an apartment and got involved with many community activities. She had a group of friends and enjoyed spending time with her family. About three years after moving into her apartment, staff at her support services agency began to notice subtle changes in her function and behavior. A physician, during a medical visit, also noticed deterioration in function and suspected early dementia related to Alzheimer's disease. Within the year she showed further decline in skills and significant physical problems, such as incontinence. She began to have crying spells and became more emotionally unstable. Shortly afterwards, she moved back to a group home as she was having a very difficult time living alone. She began to have seizures and severe swings in her sleep patterns. Her behavior and functioning deteriorated even more. She was constantly agitated and was progressively more dependent on staff for everything. Within a couple of months, she had to have 24-hour nursing care. Ms. Sudette died about 18 months after moving into the group home."*

dementia, and are affected by the same types of dementia as other people in the general population.[2,3] Such studies also show that on the average 'onset' of dementia symptoms is first observed in the late sixties and that the presentation of symptoms (such as memory loss, disorientation, language difficulties, etc.) is similar to those observed in the general population – although there are symptom variations in adults with Down syndrome (where personality changes appear more prominently). Risk factors and duration features (longevity after diagnosis) appear to be similar to those of the general population.

However, no large-scale population studies have been undertaken to validate these preliminary observations.

The National Task Group recommends increased investment into epidemiological research to examine the nature and processes of dementia among adults with intellectual and developmental disabilities, including risk factors and the incidence of the various dementias.

Recommendation #2:
Conduct studies to identify and scientifically establish the risk factors associated with the occurrence of dementia among adults with an intellectual disability.

3.0 Challenges Facing the Population

3.1 Underlying Philosophy

The National Task Group endorses the 'Edinburgh Principles'[9] which are the foundation for the design and support of community services to people with an intellectual disability affected by dementia. The Principles are structured using a four-point approach: adopting a workable philosophy of care, adapting practices at the point of service delivery, working out the coordination of diverse systems, and promoting relevant research. The Edinburgh Principles are seven statements identifying a foundation for the design and support of services to people with an intellectual disability affected by dementia as well as their caregivers. The Principles are the following:

1. Adopt an operational philosophy that promotes quality of life.

2. Affirm that individual strengths guide decision-making.

3. Involve the individual and family in all planning and services.

4. Ensure availability of appropriate diagnostic and service resources.

5. Plan and provide supports to optimize remaining in community.

6. Ensure that people with an intellectual disability have access to same dementia services provided to others in population.

7. Ensure that community dementia services' planning also involves a focus on adults with an intellectual disability.

The expectation is that the number of adults with an intellectual disability who will be affected by dementia will grow over the next 30 years. Thus, it is critical to set the underlying tone – the foundation – for the nature and character of services and supports now, so that quality services, invested in the community, will become and remain the norm during this period.

3.2 Identification, Screening, and Diagnosing

The National Task Group considers the need for early and valid identification of individuals showing signs and symptoms of cognitive impairment and dementia as an important first step in managing this age-associated challenge. The adoption of a workable and useful functional screening instrument would help staff and families establish baseline (that is, current functioning) data and would help subsequently identify and track significant changes in function that would speed referral to appropriate evaluators and diagnosticians. Follow-up assessments would help determine whether the changes being observed are linked to a disease or condition leading to dementia or may be the result of other treatable and reversible factors

3.2.1 Identification

Adopting a strategy for the tracking of trajectories of functional and cognitive decline can assist families and organizations with the formulation of a proactive approach to planning for future needs and the introduction of interventions responsive to changing needs. Surveillance strategies that identify and track indicators of cognitive decline or mild cognitive impairment, as early signs of dementia, should be instituted by any organization serving adults with intellectual disabilities, and in particular those with Down syndrome. It is also important to minimize the 'moving baseline' conundrum (when a delay in assessment results in missed measures of behavior) evident in some systems when lengthy waiting lists for assessment pose a barrier to immediate evaluation of and feedback on function. Timeliness of screenings and assessments is crucial.

As a matter of course, families and organizations should undertake efforts to gain baseline information about optimal functioning, prior to the onset of any symptoms of cognitive decline or dysfunction among individuals who may be at-risk. Such efforts might involve using paper and pencil or digitally recorded performance and function measures, which may be kept for later use. One such protocol for recording current performance is available via the website of the University of Illinois' RRTC on Aging with Developmental Disabilities.[11] Although various measures of assessing current performance are available, norms (such as those used with the Mini-Mental State Examination [MMSE] in the general population) are generally not useful because of individual variations in abilities and function associated with intellectual disability. Thus, as each individual is his or her own point of comparison (using a "personal best" approach), it is useful to periodically reapply the same measure.

It is also recognized that the degree of intellectual disability may determine the usefulness of the information derived from such measures, with such information more reliable among adults with mild or moderate limitations, and least reliable among adults with profound limitations. Notwithstanding the degree of limitation, tracking function for such early signs among adults with Down syndrome should begin after age 40. Similar tracking of adults with other etiologies of intellectual disabilities should be undertaken after adults are in their mid- to late fifties. This could be done yearly as preparation for the annual wellness visit (as provided for under the Affordable Care Act).

Screening strategies can be also tied into public awareness campaigns alerting people to early indicators of cognitive change akin to what the Alzheimer's Association utilizes via its website and publications. An informational packet for family and professional caregivers of adults with an intellectual disability could provide guidance about what types of information to collect and on how to report such screening data to physicians and other practitioners when significant changes are observed. Suggestions could be provided on how to best communicate with practitioners who may then investigate more fully such signs and symptoms.

Recommendation #3:
Develop guidelines and instructional packages for use by families and caregivers in periodically screening for signs and symptoms of dementia.

Recommendation #4:
Encourage provider agencies in the United States to implement screenings of their older-age clientele with an intellectual disability who are at-risk of or affected by dementia.

3.2.2 Screening

The National Task Group was concerned with whether there was a way that individuals with an intellectual disability could be efficiently 'screened' (and thus more accurately identified) for possible or probable signs of mild cognitive impairment and dementia (using indicators of cognitive decline or impairment and or behavioral change). It considered whether such a functional 'screening' instrument existed which could serve as a 'first instance' metric that could be tied to behavioral indicators of dementia (that is, the 'warning signs') and be able to both capture newly presented and successive changes in function. Such a 'screening' instrument would need to be constructed in a manner so it could be easily completed by direct support staff or family caregivers who had only minimal training or orientation. It could be used to document and then track changes over time using a validated scoring system. Such a system could signal when an individual should be referred for further clinical assessment. Such an instrument could also be used to track signals of mild cognitive impairment (MCI) and the possible subsequent conversion to dementia. It could also satisfy the requirement under the Affordable Care Act to detect 'cognitive impairment' as part of the health risk assessment of the annual wellness visit.

> **Why periodic assessment is useful**
>
> "As a nurse I recently assessed a 60-year old man with Down syndrome who lives in a group home for adults with Alzheimer's. The home's staff noticed that he was becoming more forgetful and that he was having difficulties with his job. His functioning was getting progressively worse so the staff asked me to come in and assess him. I found him to have profound changes – he had a lack of facial expression, apathy, and decline in the time it took him to do tasks at his work. He had no weight loss and no troubles with sleep. I suggested that he change to a different work situation that is located closer to where he lives. This meant he could walk to and from work with a minimum amount of supervision. Our next step was to compile all the information we had and take him to his primary physician for an evaluation and ask for consults for neurology and psychiatry. We also planned to have a staff meeting to come up with a plan for what we could do to help him that will enable him to remain in the group home. As his only family is a brother who lives out of town, he has had little family involvement. However, we plan to ask his brother to come to a meeting and discuss a plan A and a plan B (for when the plan A no longer works)."

The National Task Group is not proposing that a wholesale 'population screening' be undertaken where every person with an intellectual disability over a certain age is examined for symptoms of dementia. However, what is being proposed is the application of a metric (that is, a 'screen') on the behavior of individuals at-risk or suspected of evidencing changes associated with cognitive or functional decline. Such an individual assessment would help point to the need for further more extensive evaluation and measure of changes in cognitive functioning and behavior and potentially may lead to a differential diagnosis. Such screenings would also help identify those individuals with dementia-like symptoms, but whose function and behavior are being affected by other causes (such as thyroid disorders, medication interactions, depression, etc.). Such individual screenings would be best undertaken using an easily administered instrument designed to detect signal behaviors usually associated with early signs of mild cognitive impairment or dementia.

The National Task Group recognized that assessment instruments of cognitive function, many of which may include direct assessment as well as informant-based ratings, already exist and are often used by clinicians to validate suspicions and help clinically track decline or changes. Such assessment instruments, more heavily geared toward assessing memory, behavioral changes and motor and visual spatial abilities, often require their users to have extensive training and professional credentials (or are reserved for research purposes). It was noted that such instruments are useful for clinical assessment and diagnostics, but would be less likely to function as a general screening instrument – because of their complexity – that could be easily used by direct support staff or family caregivers.

The National Task Group examined a number of instruments in use for assessing behavior and indicating whether the changes are due to a disease or organic process leading to dementia. One of the instruments examined appeared to have the most utility for use by direct support staff and family caregivers as a 'first-instance' screen. This instrument was the Dementia Screening Questionnaire for Individuals with Intellectual Disabilities (DSQIID), developed by Dr. Shoumitro Deb at the University of Birmingham.[10] The DSQIID, which is used in various parts of the world, is an informant-based instrument which would enable agencies to record change in behaviors typically noted as indicators or warning signs for dementia, and also might be useful for the cognitive impairment aspect of the annual wellness visit screening enabled by the Affordable Care Act. The National Task Group noted the potential utility of the DSQIID to serve as a 'first-instance' screening instrument – as it can be easily used by direct support staff and family caregivers and allows for scoring in a manner which would provide a metric to gauge change over time.

In the United States, an adapted version of the DSQIID (termed the "Dementia Screening Tool") is currently being used by a project in the greater Philadelphia area with Pennhurst class members, as well as with another group of older adults, as part of their mandated annual review. The project, under the auspice of Philadelphia Coordinated Health Care, has piloted the adapted version and has expressed an interest in working with the National Task Group to create an applicable variant of this instrument and expand its application to other parts of the United States.

Recommendation #5:
Examine the utility of adopting an instrument such as an adapted Dementia Screening Questionnaire for Individuals with Intellectual Disabilities for use in preparation for the annual wellness visit.

Recommendation #6:
Conduct an evaluation of a workable scoring scheme for the Dementia Screening Questionnaire for Individuals with Intellectual Disabilities that would help identify individuals in decline.

3.2.3 Diagnosing

Since currently Alzheimer's and some other forms of neurodegenerative disorders can be only proven on autopsy, a diagnosis is made clinically only after reversible causes and co-morbid confounding variables have been eliminated by laboratory workups and neuroimaging[8]. Reviewing an individual's premorbid activities of daily living, that show evidence of a progressive loss of function, is essential to the diagnosis. Age-related functional decline will continue to be a confounding element to the diagnosis. Biomarkers may eventually help confirm the diagnosis when they become available, beyond current usage in research studies.

The process of diagnosing also includes clinical observation for symptoms of mild cognitive impairment (MCI). Currently research shows that in some instances of MCI, there is a progression to early stage dementia. Dr. Wayne Silverman at the Kennedy Krieger Institute in Baltimore, Maryland and his colleagues have shown that among adults with Down syndrome there is a 33% conversion rate from MCI to early dementia within an 18-month period.[12] Given these findings, it is prudent to consider capturing early signals of functional change that may indicate the presence of MCI as well as looking for the early indicators of dementia. Also, as it is important to note that not all instances of MCI will lead to early dementia, early differential assessments can help with symptom remediation.

Formal diagnostics are the domain of specialists generally found in more urbanized settings. The close proximity of a regional diagnostic center can provide valuable support with early diagnosis and ongoing evaluations. Access to such specialized services outside major centers is, however, a barrier. There is a potential role for technology to be used to better support individuals, agencies, and professionals living in smaller cities, towns, and rural areas. Diagnostic resources also need to be reliable and accurate, as some instances of symptom presentation and change may be attributable to other causes, and not to dementia. Often adults with an intellectual disability, particularly those with multiple complications or severe intellectual impairment, may be misdiagnosed and the course of treatment consequently misdirected. Loss of function, personality and emotional changes, and loss of activity may be the result of other diseases or physical or sensory problems (such as nutritional deficiencies, thyroid abnormalities, or hearing impairments). Having well-trained and reliable diagnostic resources available to aid in assessing adults with an intellectual disability is crucial to accurate diagnosing and appropriate remedial treatment.

One area that warrants further development is how to expand both the number and expertise of practitioners qualified to conduct diagnostic work-ups. Many community practitioners are highly qualified to diagnose dementia in adults in the general population, but may be stymied when confronted by an adult with a lifelong intellectual disability. Conversely, practitioners skilled in assessing adults with an intellectual disability may be unfamiliar with neurodegenerative conditions associated with older age. One model that may have promise is short-term clinical placements of generic practitioners in regional intellectual disabilities' diagnostic centers that assess geriatric problems in older adults with an intellectual disability. Another may be specialized continuing education sponsored by regional dementia diagnostic centers which may contract with intellectual disabilities specialists to do general trainings. Other training models, such as the use of practicums and webinars, also have promise.

Recommendation #7:
Establish undergraduate, graduate, and continuing education programs, using various modalities, to enhance the diagnostic skills of community practitioners.

Recommendation #8:
Promote the exchange of information among clinicians regarding technical aspects of existing assessment and diagnostic instruments for confirming presence of dementia in people with an intellectual disability.

3.3 Dementia and Family Caregiving

3.3.1 Impact

As a group, family caregivers represent committed parents, siblings, the children of siblings, and other extended family members or significant others. As families, they've made decisions to have their relative with an intellectual disability live with them or to provide ongoing familial support when their relative resided independent from them. However, these supports may be compromised with the presence of dementia, which has a tremendous impact not only on the individual but also on the family, as a whole, via familial relations and supports.

In the United States, it has been estimated that some 75 percent of older-aged adults with a developmental disability reside with their families. With increasing age, age-associated impairments and pathologies begin to take prominence and are amongst the challenges facing both family caregivers and adults with an intellectual disability. Information derived from the *State of State* report for 2010, issued by Professor David Braddock and his colleagues at the University of Colorado's Coleman Institute for Cognitive Disabilities, indicates that there are approximately 2.88 million individuals of all ages with developmental disabilities residing with their families.[6] Of these, approximately 731,000 adults reside with family caregivers age 60+ and another 1.0 million reside with family caregivers who are between age 41 and 59. Using the age of onset data (that the average age of onset of dementia for adults with Down syndrome is about 52 and the average age of onset of dementia for adults with other

> ***Often changes in behavior are subtle and the demands on the primary caregiver manifold***
>
> *Mrs. Roberta Wilson is in her 70s and has been taking care of her daughter with Down syndrome since her daughter was born. She lives with her husband and a niece in a home they have had since they were young. Her daughter has always lived with her and has never been registered with the local social services agency. Lately, her husband, a veteran, has begun to have difficulty walking and had to leave his job at the local lumber mill. He has been referred to a near-by Veteran's Affairs hospital for assessment. Mrs. Wilson's daughter, now in her 50s, has always been helpful to her and the neighbors. Within the past year, Mrs. Wilson has noticed that her daughter has been lethargic, not sleeping well, and is beginning to have difficulties doing her normal daily things at the house. Mrs. Wilson's neighbor also told her she suspects that her daughter is not the same. She is concerned that her daughter may be having some kind of problem. Her niece, who works for a local doctor, told her about a local clinic for adults with intellectual disabilities and suggested she contact them and ask if they could help her. Mrs. Wilson is concerned that with her husband's condition and her daughter's new problems, she will no longer be able to cope with caring for both.*

intellectual disabilities is about 67) then the impact on older caregivers will be considerable. This may be especially so since it is estimated that about 33,000 adults currently living at home with older family caregivers might be affected by mild cognitive impairment (MCI) or dementia.

Because of specific factors related to Down syndrome (such as accelerated aging) many adults who may be a risk of dementia are still living with parents who are in their younger-older years. Thus, the greatest impact on families may come when parents are in the late sixties or early seventies. This impact may differ among families caring for adults with an intellectual disability of other etiologies, as they may be older when the first symptoms of dementia begin to appear. In such situations, parents may no longer be the caregivers. Siblings, other relatives, and housemates may be the ones impacted as they often assume the responsibility of care.

One of the realities of how dementia affects people is that eventually self-direction and self-care abilities are lost and an affected person can no longer live by him or herself. This same situation is true for adults with an intellectual disability who may have been successfully independent. There are a number of early changes that signal the failing ability to remain in one's own home. In addition, this is also true for adults still living with caregivers. Research has shown that the types of initial changes in the individual observed by caregivers include forget-fulness, losses in personal skills, and changes in personality.[13] Also, challenging behaviors, such as wandering, becoming argumentative or aggressive, and problems, such as experiencing sleeping difficulties, incontinence, and rummaging – generally indicative of mid-stage dementia – are usually identified early by family caregivers. The supervision required will increase due to the changes in function.

As dementia leads to progressive losses in memory, self-care abilities, judgment, and eventually to a loss of mobility, someone affected by dementia from the mid-stage on cannot be left alone and unsupervised. When this happens the burden of support falls upon a family member or someone else who has the responsibility for supervision and care.

> **How provider agencies are affected by the increase of dementia among their clientele**
>
> "I am the director of a residential program serving 155 adults with intellectual and developmental disabilities, many of whom have lived in our group homes for over 20 years. When we were planning for making environments accessible by incorporating universal design principles, we were shocked at the rate of rapid onset of aging and dementia symptoms among the residents throughout our service area. We struggled to make resources available, such as shifting staff from one location to another as adults were diagnosed with dementia. In addition to the staffing changes, we also enabled our nurses and behavior analysts to be more mobile so they could more effectively support the group home residents and staff. We supplied them with laptops and smart-phones to enhance communication and we added specialized staff to function as liaisons among community physicians, our clinical staff, and our residential team. In the last year we have seen medical appointments increase from some 4,000 to 4,500. We have experienced a dramatic increase in protocols in the areas of direct support, assistive devices, swallowing and choking precautions, and gait problems. It has been extremely challenging to support these adults as they 'aged in place'. We had to ask some adults to move as less than 50% of our homes are accessible and our inventory of homes that appropriately support someone with dementia and complex medical issues is limited. In some homes, we had to move some of the adults to first floor bedrooms and enhance staff training to support individuals with swallowing and chewing problems, and losses of ambulation. We've been helped by hospice with providing end-of-life supports."

Recommendation #9: Conduct studies on the impact of aging of family caregivers on the support and care of adults with intellectual disabilities residing in at-home settings.

3.3.2 Caregiving at Home

Currently, the numbers of older family caregivers still providing home-based supports and supervision of an adult with an intellectual disability are significant and the relief of costs that would otherwise be borne by governments quite staggering. With the onset of dementia these caregivers will be taxed to continue to provide home-based care, and, like their peers in the general population caring for a spouse or parent (without an intellectual disability) affected by dementia, will look for ways to accommodate the changes resulting from dementia, as well as to continue to provide care while seeking outside assistance and support. Family caregivers living separately from their relative with an intellectual disability, who is beginning to be affected by dementia, also struggle to understand the disease, maintain their relationship, and ensure that appropriate quality supports are provided in a timely fashion.

Frequently, at the very time when older family caregivers (usually parents) of adults with an intellectual disability are trying to reduce their life-long caregiving responsibilities for a son or daughter living at home, and are anticipating that perhaps their son or daughter may be moving on to another stage of their life, such caregivers are potentially facing a renewed demand for extensive supervision and self-care supports when confronted with the onset of dementia. Siblings and other relatives who are caregivers may also be confronted with profound changes in their relative with an intellectual disability and struggle with maintaining longtime familial support alongside their other commitments such as work, raising their own children, or caring for older-aged parents. This phenomenon is faced daily by the thousands of spouses and other relatives in the general population who are now faced with providing (often unaided) support and care to a relative affected by dementia. Although most life-long family caregivers of adults with an intellectual disability appear to have more capacity to adapt to the day-to-day responsibilities of caregiving, eventually even for them the growing responsibility of care will be overwhelming.

Some families experience and have coped with a lifetime level of care that is functional, but the challenges of providing the add-on care for dementia becomes problematic and can often exceed their capacities. Caregiver's physical and emotional health as well as family finances can often be compromised and stretched beyond capacities to continue to cope and care at home for someone affected by dementia. When the demands of caregiving exceed a family's abilities, many families tend to seek community care alternatives that may mirror their home life experiences. For some, other family members may agree to provide aid as caregiving becomes problematic or these caregivers may turn to agencies that are operating and promoting alternative care settings, primarily small neighborhood-based group homes. Many seek an alternative family or group home environment as the residence of choice for their relative when they decide they can no longer offer care in the family home.

Overall, families will be facing increased demands as they continue to provide supports for adult relatives living at home. More needs to be undertaken to assist families caught up in situations of continued caregiving for adults with diminishing abilities – both to preclude institutional referrals and to support them with continued caregiving and enhance their quality of life. Notwithstanding the current fiscal climate in the country, where in many states' services are being curtailed and eliminated, the area of family supports requires continued emphasis. States need to be cognizant that for each person still living at home, the demand for a publicly supported living situation remains unannounced. In addition, caregivers should be able to look for supports from the National Caregiver Support Program.

In many cases, a personal desire to keep 'the issue at home' reflects the family's cultural values or religious beliefs. To assist families in such situations, organizations need to be sensitive to these values and beliefs and organize supports with this in mind. States also need to account for cultural differences when planning for, and implementing family support for dementia care directed towards adults with intellectual disabilities and diverse demographic (e.g., cultural, ritual) backgrounds.

Recommendation #10: Enhance family support services to include efforts to help caregivers to identify and receive assistance for aiding adults with an intellectual disability affected by dementia.

3.4 Community Providers and Community Living

3.4.1 Impact

The National Task Group recognized that although some adults with an intellectual disability will have dependencies and live with families or in supported housing, most may be sustaining themselves, functioning independently, and living normal lives within our communities. However, as they age, some of these 'unknown' adults may be at-risk for age-associated problems (including dementia). Such situations may be particularly problematic as their decline may go unnoticed or they may have difficulties with obtaining supports from public or private agencies. They may have lived most of their lives without formal connections to public or private organizations, but at times have received substantial support from their parents, other relatives or friends. When disconnected from formal services, the identification and subsequent follow-up of the onset of dementia is more problematic. The lack of a documented history of formal contact or previous eligibility for services can pose a problem when attempting to access aging-related services. One potential solution to this problem is the use of 'presumptive eligibility' which would recognize anecdotal information and other 'loose' evidence of marginalized status (e.g., childhood problems, no formal schooling, sketchy work history, community recognition of history of marginalization, lifelong impairment) and thus open up opportunities for the receipt of services related to their current situation.

> **As one agency administrator noted**...
>
> *"A big challenge we see is the impact the person with dementia has on housemates. Their ability to understand what is happening with the person varies widely. Even the most caring housemate can become exhausted by the changes occurring and lose tolerance. We continually assess the pros and cons of designating residential sites specifically for people with dementia. Many of the people we serve have lived with the same housemates for many years and have established relationships. While we see the benefits of serving people with similar needs in one setting (specialized staff training, enhanced nursing oversight, fewer transitions, etc.), we are concerned that moving a person who has dementia to a completely new setting may speed up the disease progression and take away the remaining connections he/she has. Our approach thus far has been to serve people in their home for as long as possible, adding staff support and training as well as working with housemates to alleviate their distress. However, most people have increasing health risks that eventually exceed the capacity of our current non-medical community-based settings. We have begun exploring the feasibility of developing a dementia-oriented group home."*

With respect to community long-term care settings, there are a variety of community residential configurations in which adults with an intellectual disability may live. These include such situations as living alone or with a long term partner with some form of support from informal or formal services or living in various community supported settings with support from formal organizations. In any of these situations the onset of dementia may present a challenge for the persons who are providing supports and certainly may lead to a crisis if the symptoms of dementia become problematic and or as they progressively compromise independent functioning.

The National Association of State Directors of Developmental Disabilities Services has reported that dementia was noted in about 1.5% of adults in formal services in states and regions participating in the National Core Indicators Project.[14] Of this 1.5%, some 35% were adults with Down syndrome and about 70% lived in some type of formal residential setting (primarily group homes). In terms of functional levels, these adults were equally distributed among the mild, moderate, and severe IQ categories. Although these data indicate that known adults in services do represent a significant number of individuals and that those at-risk of or with dementia are a number that is gaining attention, they also possibly represent a severe under-count or under-diagnosis that may be present in many services. This apparent under-count may reflect less formal connectedness with older adults still living in the community on their own or with families. It may also be indicative of under-diagnosis – that is, a lack of awareness to early indicators of functional decline or of early dementia.

Recommendation #11: Conduct nationwide medico-economic studies on the financial impact of dementia among people with intellectual disabilities in various service provision settings.

3.4.2 Community Living Support Settings

The National Task Group considered a number of scenarios where an adult with an intellectual disability might need assistance and potentially require greater levels of support once dementia symptoms become pronounced. These could include any of the following:

- *Adults who live alone or with a housemate.* In such instances when mild cognitive Impairment might be suspected or early stage dementia may be present, such a situation may remain static as long as the early aspects of dementia do not jeopardize the person's safety or the person can still function with some degree of independence. In such situations, however, there would be benefits from outreach, support such as visitations, and personal support arrangements. It is critical to consider and plan for an alternative living arrangement that can provide the personal assistance and supervision required when dementia progresses from early to later stages. If the person is unknown to any formal services, recognizing a need for and gaining access to specialized dementia supports may be difficult.

- *Adults who live in a group home or apartment.* At times it is prudent to move an adult to a formal community residential setting, such as a group home or a cluster apartment. When a commitment is made to enable the person to continue to live in this setting, it is referred to as 'aging-in-place'. Such situations exist when an agency is operating a community group home, with possibly one or two adults affected by dementia, and decides to provide long-term, in-home supportive services for the person(s) affected. In these instances dementia-related care, directed toward supporting a particular individual, is usually modified as the person experiences progressive decline. Further, individual supports are adapted and modified at each stage of dementia to permit the individual to remain within that home for as long as possible (often until death).

- *Adults who live in a specialized 'dementia-capable' residence.* Some agencies maintain *a home or multiple homes* which specialize in dementia care. Such homes generally use an "in-place progression" model. This model usually involves a cluster of persons with varying levels of dementia residing in one or more specialty group homes. The individuals may be moved between the homes contingent on the level of support needed. Alternatively, staffing and other supports may vary, permitting the individuals to remain in place. With sufficient numbers of people affected by dementia, there may be multiple homes providing stage-related levels of supports. These homes are usually staffed by caregivers who have received specialty training and the physical environment is adapted or designed to accommodate progressive decline in physical and mental abilities.

Recommendation #12: Plan for and develop more specialized group homes for dementia care as well as develop support capacities for helping adults affected by dementia still living on their own or with their family.

3.4.3 Support Services

The National Task Group recognized the impact that dementia may have on persons who are residing with, are friends of, or are providing supports to adults affected by dementia. These could include any of the following situations:

• There may be instances where living with individuals affected by dementia will have an impact on housemates. In these instances there is a need to balance the needs of all individuals as this may lead to a highly stressful situation resulting in exhaustion of housemates. There is a need to critically assess the impact of dementia on others who are close to the individual, and consider living compatibility re-assessments, when appropriate.

• There may be instances where supportive information explaining the nature of dementia in an elementary way can help housemates or relatives understand what is happening and what will happen to their friend or relative (e.g. materials explaining dementia, changes, prognosis). There is a need for more educational materials to help people with an intellectual disability understand dementia and its consequences.

• There may be instances when the individual may move from his or her current residence to an alternative living arrangement. In these instances, former housemates may want to maintain contact with the individual. They may also want to voice and/or resolve their feelings about past interactions that may have resulted in lingering guilt, sadness, confusion and loss. There is a need for a variety of psycho-social supports for persons affected by someone they know with dementia.

• There may be instances when there is the intention to support the person with dementia within his or her home until the end of life. In these instances, housemates should have access to the necessary resources that will help them cope with emotional aspects of death as well as the physical aspects of death. There is a need for close coordination with hospice or other end-of-life support organizations.

• There may be instances when supportive education and training should be made available to housemates, relatives, and staff (e.g., materials to explain dementia, changes in function, understanding grief and loss for persons with cognitive impairment, staff burn-out, and "compassion fatigue"). There is a need to produce and disseminate appropriate written educational materials.

The National Task Group also acknowledges that in dementia care settings the philosophy of support provision has by necessity changed from one with a focus on measurable goals of learning new skills to one with a focus of maintaining existing skills and any new gains, as well as more 'hands-on' activities and supportive care. It also recognizes that with respect to dementia-capable care, the focus is on enabling continued capabilities to the greatest extent possible, that expectations of new learning may need to be tempered, and over time the loss of abilities supported.

With respect to supports for early identification and potential re-assessment of life plans, the National Task Group proposes that local agencies initiate screenings or assessments of their current "at-risk" clientele to determine which individuals may be showing signs of mild cognitive impairment or signs of early dementia and that they develop a range of dementia-capable services to help those adults who are affected as well as their families. To this end, local provider agencies should consider:

● Conducting outreach via generic aging providers (e.g., senior centers, geriatric health clinics, long term care providers) to identify adults with an intellectual disability who may be showing symptoms of dementia.

● Investing in screening and assessment resources or establishing links to services available in their community that do such assessments.

● Adapting their residential services, using a group home model, for long-term care and support of adults aging with an intellectual disability and who may become affected by dementia.

● Commissioning specialized clinical consultation teams (of physicians, nurses, psychologists, social welfare workers and adapted environment specialists) to aid family and staff caregivers to better understand and provide care for and supports to adults affected by dementia.

● Arranging for medical and mental health assessments and follow-up to attend to the physical and mental health needs of adults affected by dementia.

Recommendation #13: Plan and develop community-based 'dementia-capable' supports to address the needs of those persons at-risk or affected by dementia.

Recommendation #14: Develop and disseminate social care practice guidelines to community agencies and professionals that address assessment, service development and life planning for adults with an intellectual disability presenting with symptoms of dementia.

3.5 Health and Secondary Conditions Affecting Adults with Dementia

The National Task Group noted that there was a need to emphasize increased vulnerability to decline from secondary or other conditions, and increased follow-up and supervision to decrease emergency room usage and visits. Often among adults with an intellectual disability, there is concern about the predisposition to attribute many complications after diagnosis to Alzheimer's disease, when, in fact, a differential diagnosis may identify other causes of sudden decline (e.g., metabolic derangements, medication side effects, drug-drug interactions, etc.).

3.5.1 Evaluations

The approach to evaluating an adult with an intellectual disability and suspected memory loss is highly variable, depending largely on the organization supporting the individual as well as the provider doing the assessment.   General practitioners who treat adults with an intellectual disability generally received little to no specialized training on the specific needs of such adults who are aging. The likelihood is quite small that they have any specialized training or basis of experience in investigating cognitive impairment in an adult with an intellectual disability. Further, access to specialists within this field is quite limited. There is also no single accepted 'gold standard' instrument for screening or assessing memory in individuals with an intellectual disability, which further leads to the disparate approaches to an office evaluation.

Specialty assessments typically take place in designated memory clinics, or by geriatricians, psychologists, neurologists, or psychiatrists who have experience and expertise in evaluating adults with an intellectual disability. However, even among specialists in this field, there is no singular systematic approach to an assessment. History is the cornerstone of a dementia diagnosis. As Alzheimer's disease and other dementing illnesses are clinical diagnoses of both inclusion and exclusion, it is important that a thorough and comprehensive history be conducted to assess for patterns consistent with a developing dementia and to uncover other features that might suggest other underlying causes or contributing factors.

When done appropriately, such assessments can be a somewhat labor-intensive process, which can be a fundamental hindrance to such a task being undertaken within the constraints of a brief office encounter. This barrier could be overcome to some extent through more systematic longitudinal history-taking and use of an easily-usable screening instrument or checklist by caregivers and families prior to office appointments.

3.5.2 Therapeutics

Medications. After the diagnosis of dementia has been made it is the role of the healthcare practitioner to consider pharmacological and non-pharmacological therapeutic measure to offset the disease process. Drugs such as donepezil, a reversible ACh-E inhibitor, may slow the progression of dementia. Medical treatment of associated vascular risk factors and compliance with diet, exercise, and nutrition needs to be adhered to. Challenging behaviors such as agitation, wandering, aggression, anxiety, depression and psychotic behavior need to be carefully monitored and evaluated. Side effects to medications can often masquerade as cognitive and behavioral complications of dementia. In addition, the development of new problems, such as seizures, often complicates the plan of care.

Diet and nutrition. Diet and nutrition is an important determinant to health and meets not only the basic biological or physiological needs but also emotional, psychological, social, and cultural needs that profoundly affects the quality of life of all individuals. The effects of dementia, already noted previously, superimposed on the effects of an intellectual disability on older persons, lead to complex nutrition-related issues that require the services of competent food and nutrition professionals for proper nutrition screening, assessment, intervention, evaluation of outcomes, and continual monitoring of the individuals' nutritional status. These professionals should be an integral part of a collaborative health care team that provides coordinated services to achieve jointly determined health outcomes. They need to work within a team in providing individualized nutrition therapy for obesity, diabetes, cardiovascular diseases, gastrointestinal problems (such as gastroesophageal reflux and constipation), renal disease, musculoskeletal disorders (such as osteoporosis), and nutrition support for individuals who have difficulty swallowing or who need tube feeding as an alternate nutrition route.

These professionals can also translate abstract nutrition principles and guidelines into healthful food choices that are customized to the individuals' personal preferences, socio-economic and cultural factors, and physio-logical mechanics of food intake. In addition, they can give sound information regarding the use of dietary supplements, functional foods, nutraceuticals and herbal therapies. Unaware of the potential harm caused by improper use of these products, some family caregivers may be using some of these products themselves for perceived health benefits and may be giving them also to their family members with an intellectual disability and dementia.

> **A clinic worker told us about a 69-year-old woman with cerebral palsy and intellectual disability experiencing decline**
>
> *"Ms. Williams had been referred to the clinic as she had declined any further participation in her day program or going out for recreational outings, while spending most of her time alone in her room. She appeared to have lost her ability to verbally communicate and cried frequently. The staff at her group home noticed this and reported it to her nurse who made an appointment with a physician. The physician diagnosed her cognitive and functioning losses as 'dementia' due to her pre-existing intellectual disability. The nurse was not convinced of this diagnosis and made an appointment with our clinic, which specializes in assessment and intervention for adults with memory and cognitive loss. The clinical specialist diagnosed Ms. Williams with severe arthritis and gastrointestinal reflux. His treatment focus was on pain management and a reduction of discomfort due to the reflux. Within a week of the assessment and pharmaceutical interventions, Ms. Williams began eating again, attending her day program, and participating in her favorite recreational activities."*

Oral health, eating, and swallowing. Among other factors, poor oral health, chewing and swallowing problems, appetite and taste changes, decline in functional status, anxiety and confusion, behavioral and neurological problems, and certain medications can reduce food intake, cause weight loss and frailty, and adversely affect nutritional status. Good nutritional status can help maintain, as much as possible, the health of individuals at various stages of dementia. Various tools are available for screening individuals for swallowing capabilities and risk factors of malnutrition. However, signs such as poor food intake or nutrient retention (which may be due to various factors), weight loss, and decline in functional status should alert caregivers for a prompt referral to qualified health professionals.

Psychological and psychiatric evaluations. Consideration of psychological and/or psychiatric evaluations should always be entertained when there are questions and concerns that the challenging behaviors have become overly problematic. The goals of therapy and the means by which an outcome is to be assessed needs to be delineated up front, as well should be the expectations of therapy. The duration of therapy should also be considered when observing for a desired effect. The success and value of a healthcare plan will be much more positive if it is conducted with clarity and there is communication with families, direct support professionals, nurses, support agencies, and other healthcare professionals who also are working together with the individual with dementia.

Recommendation #15: Develop and disseminate a set of nutritional and dietary guidelines appropriate for persons with an intellectual disability affected by dementia.

3.5.3 Medical reviews

When the individual begins to exhibit changes consistent with mid-stage disease, anticipatory guidance should include a focus both on the 'here and now' as well as future goals including care, living arrangements and advance directives. Initiating this process in early or mid-stage disease is beneficial in that it allows for the individuals affected and their caregiver(s) to consider scenarios that may be encountered in late-stage disease, while avoiding making decisions in a crisis situation.

Components of this discussion should include the following:

• How overall goals of care can change as dementia progresses, exploring the options of changing goals to focus primarily on comfort, avoidance of hospitalization, and other invasive means of evaluation and treatment and that notwithstanding, reversible medical conditions should not be ignored, always entertained, and promptly and appropriately managed.

• Assessment for pre-existing beliefs or decisions previously made about end-of-life care.

• Assessment of need for fully accessible living situations.

• A nutritional assessment to include dietary and hydration recommendations.

• The occurrence of challenging behaviors and the provision of staff training in anticipation of such changes.

• The expectation of progressive gait dysfunction that can lead to falls and possible trauma and the need for fully accessible living settings.

• Review the features of late stage disease.

• The possibility of progressive dysphagia with associated aspiration and aspiration pneumonia that could prompt future discussion regarding wishes for feeding tube placement as an alternate way of getting nutrition.

• The question of cardiopulmonary resuscitation, explaining all unfamiliar terms including "full code", "do-not-attempt resuscitation" (DNAR), do-not-resuscitate (DNR), and do-not-intubate (DNI).

• Assessing wishes for resuscitation efforts in the setting of advanced dementia.

• Consideration of palliative care as individuals are often unable to express issues related to pain and discomfort.

• An outline of the features of advanced dementia that comprise criteria for hospice eligibility criteria such as frailty and advanced debility.

• The option of future enrollment in hospice services and determination if goals of care would be in alignment with hospice services.

3.5.4 Health Disparities

The National Task Group concurs with findings of the U.S. Surgeon General that disparities existent in the United States with respect to access to health services and, more importantly, recognizes the dearth of training and experienced medical professionals and clinicians capable of assessing and diagnosing dementia in adults with an intellectual disability.[15]

The nature of these inequities often fit what Dr. Gloria Krahn of the US Centers for Disease Control and Prevention has termed a 'cascade of disparities', where there is a higher prevalence of adverse conditions, inadequate attention to care needs, inadequate focus on health promotion, and inadequate access to quality health care services.[16] To begin to address these inequities, the National Task Group recognizes the need for more functional information to direct service providers, primary care physicians, and other health professionals about the health needs of older adults with an intellectual disability affected by dementia as well as specific protocols for treatment.

Studies [14, 17] are beginning to show data on the health status characteristics of adults with an intellectual disability and dementia, as well as how they utilize services. Preliminary data from the Longitudinal Health and Intellectual Disability Study (LHIDS) at the University of Illinois at Chicago indicate that many adults with dementia have several co-morbid health conditions, including high blood cholesterol, thyroid disorders, diabetes, cardiovascular, gastrointestinal and genital-urinary problems. They also have psychiatric conditions, including depression and anxiety, as well as sleep disorders and seizures.

The National Core Indicators data[14] show that with respect to health services utilization, adults with dementia who are known to state agencies are high users of health services and receive a high number of health screenings. It is important to point out that these data come from *known individuals* who are associated with provider agencies and in large part are in formal residential care. To what degree other adults with an intellectual disability and dementia would mirror these characteristics is unknown; however, as most adults distant from formal services are low users of health services, it would be expected that such low utilization would also be true for adults with dementia living on their own or with families.

Given such preliminary data on the health and health services usage of adults with an intellectual disability and dementia, it is doubly important to bridge the gap in health practitioners' understanding of prevalent co-morbid conditions found among this group. The use of medical screens for some of these coincident conditions, during physician visits, would be strongly indicated. It is also important to have more population-level information on the health factors associated with dementia among adults with an intellectual disability and understand better which conditions or diseases may be precursors to, or associated with, the onset of the symptoms of dementia.

With this in mind, the National Task Group is recommending that the American Academy of Developmental Medicine and Dentistry, in consultation with other groups, prepare and issue a set of health care practice guidelines, directed toward primary care physicians and other health professionals, with the goal of decreasing the disparities in this area of health provision and increasing competencies in providing appropriate health care to this population. The National Task Group is also recommending a more extensive research effort on the health conditions associated with dementia in adults with an intellectual disability to provide an empirical basis for the guidelines.

In addition, the National Task Group recognizes that often it is difficult to obtain consent for assessment and interventions for adults with an intellectual disability with suspected dementia. These difficulties stem from challenges in communicating symptoms of decline, increased use of medications over a lifetime (exacerbating side effects of medications), masking or mimicking of symptoms by other diseases, and expectations by health care providers of automatic decline in adults with an intellectual disability. It is important that informal and formal caregivers are trained in skills of health care advocacy to obtain consent that would facilitate quality assessments and interventions.

Due to the complexity of assessment, diagnosis, and interventions with adults with an intellectual disability who are suspected of having dementia, the National Task Group recognizes the importance of developing practice guidelines that cover health care advocacy. Such guidelines should be designed in a manner so as to help family caregivers and agency staff enhance their health care advocacy skills and reduce barriers to health services access.

Recommendation #16: Develop and disseminate health practice guidelines to aid primary care physicians and related health practitioners address assessment and follow-up treatment of adults with an intellectual disability presenting with symptoms of dementia.

Recommendation #17: Conduct studies on the nature and extent of health compromises, conditions, and diseases found among adults with an intellectual disability and affected by dementia.

4.0 Community Services

4.1 Enabling Quality of Life and Community Living

The National Task Group is committed to promoting care and supports of adults with an intellectual disability affected by dementia in as much as possible in community settings. The National Task Group recognized that in most instances continued community living is viable and warranted both from a human rights perspective and good care practices. [18] With appropriate supports and supervision most, if not all, adults with an intellectual disability can continue to reside in some community living setting and enjoy an enhanced quality of life.

The National Task Group recognized that dementia is generally progressive and can present with special needs and over time impose increasing demands for personal care and special supports. The National Task Group contends that what is generally needed by adults with an intellectual disability affected by dementia is a safe place to live (that is, appropriate housing) with supervision and staff or family caregiver support. Also needed is a daily regime that provides for purposeful engagement based on individual needs but is organized so as not to cause anxiety and confusion, medical follow-up for treatment of the course of the dementia, and surveillance and management of co-morbid conditions. With progression, specialized care in the later stages (including hospice and nursing supports) is needed; as is a continued connection with familiar and supportive memories as well as friends, family and others. [18] If an adult is connected to a spiritual care community, this relationship might be continued and fostered.

Researchers at the University of Illinois at Chicago have shown that international models of more personalized dementia-capable care are moving away from referrals for institutional admissions (in long-term care or aged care facilities) and toward care in small, specialized group homes in the community.[19] These models are becoming more prevalent among agencies that recognize specialized dementia care as a necessity and have the resources to devote one or more small homes for adults affected by dementia. Some agencies have created a cluster of purpose-built group homes (located in a typical neighborhood setting) designed specifically for dementia supports. Most agencies, however, still use an 'aging-in-place' model, on an individual basis, where affected adults can remain in their home for as long as their needs can be safely and adequately met.

The National Task Group also recognizes that many adults with an intellectual disability affected by dementia still live with their aging parents who continue to provide for their needs. Their needs are like any other adults affected by dementia. The main difference is that persons with an intellectual disability will also be affected by their lifelong impairment and may already have compromised function. Family needs, however, will vary. Many older-aged parents and siblings grow less able to provide day-to-day care and, with their relative's progressively diminishing capabilities, they are doubly challenged.

> **One agency clinician told us...**
>
> "We have developed an approach of trying to keep the person in their home for as long as humanly and safely possible. I also go out to each of the houses that have someone with symptoms of dementia and provide education for both the staff and the residents who are living there. It's kind of a support group for the residents and they really seem to benefit. We have a booklet on dementia that is written for residents to understand. I also have a model of a brain that shows what happens with dementia. The residents seem to like that as well. I think that through just identifying with the other residents – that it is difficult to live with someone who is going through these things – there is benefit. I think it helps them to know that we are all working through this together."

The physical environments in private homes and organizations providing residential and program supports for people with an intellectual disability and dementia are often not designed to decrease confusion and support remaining skills. However, simple and inexpensive changes in the environment can be made. Consequently, the National Task Group recommends providing training on environmental design and supports to enable living and program environments to be optimized for adults with an intellectual disability impaired by dementia.

Adherence to quality of care and quality of life principles includes acknowledging that transitions to alternative living circumstances of an individual affected by dementia should reflect choices, the person-centered best interest of the individual, the need to provide levels of care not possible in the current setting, and adherence to the notion of the least restrictive appropriate alternative.[20] In this vein, the National Task Group believes that critical aspects of maintaining a high quality of life and supporting people to remain in their homes and communities include:

• Proactive planning by families, organizations and service providers.

• On-going information and support to families.

• Staff trained on aging and dementia care with access to ongoing training and support related to individual physical, behavioral and psychological symptoms that may occur over the course of dementia.

• Sensitivity to staff and family caregivers who may experience compassion fatigue, stress, and caregiver burden.

• Modifications to housing and other physical environments used by persons with dementia.

• Adaptations to activities and flexibility by program services to accommodate changes due to dementia.

• Use of and collaboration between existing aging, dementia, and intellectual disabilities networks, and a sharing of expertise, experience and resources.

• Cooperation across policy-maker and provider systems that enhances flexibility to meet the variable and changing needs of the individual.

• Emphasis on the maintenance of acquired and existing skills over unrealistic habilitative goals of acquiring new skills.

4.2 New Directions

The National Task Group believes that organizations that provide support to persons with Alzheimer's disease should strive to align themselves to meet the needs of individuals with an intellectual disability as they experience changes as a result of dementia. Such organizations also need to be sensitive to the needs of persons with an intellectual disability and dementia who begin to experience the progression of the disease.

The National Task Group believes that community organizations and agencies, whether affiliated with the intellectual disabilities or with generic Alzheimer's disease and related dementias network need to consider how to broaden their influence and impact among families, caregivers, advocates, and persons with dementia. It is recognized that a variety of concerns and dilemmas exist. Some of these concerns and dilemmas encountered by community organizations include:

• Disseminating information to a variety of audiences, including ways of sharing information and diagnosis with people with an intellectual disability.

• Providing a detailed knowledge-base of Alzheimer's disease and other dementias (for instance, practitioners, support staff and family members who need to understand the stage progression of the disease).

• Developing strong and visible 'working-based' connections with the local area agencies on aging (AAAs).

• Cross-training between the aging and intellectual disabilities networks, and building collaborative partnerships with intellectual disabilities, aging and dementia organizations and services.

• Developing strategies to reduce staff turnover to provide continuity of care.

• Increasing the availability of diagnostic and clinical supports.

• Increasing the availability of appropriate day services and day programming.

• Coordinating the provision of appropriate residential options for persons with dementia.

• Planning and preparing for future needs.

• Ensuring the availability of palliative and hospice care in community and home settings

Many states have initiated special efforts to address the growing numbers of clientele in both public and private sector services affected by dementia. Some have authorized and are funding specialized group homes or other housing that is 'dementia capable'. Others have instituted family support programs to help caregivers facing home-based dementia-related care demands and have instituted training and education programs for provider agencies and families. However, these are not wholesale efforts and much that is needed by the provider sector in terms of additional financial and programmatic supports remains elusive. Most states do not have statistics on the number of adults affected or a catalogue of dementia-related programs and services, and many have not yet undertaken any special efforts in this area.

The National Task Group proposes that state developmental disabilities agencies and their constituent organizations recognize the growing need for special planning and attention to dementia within their clientele, and develop and support appropriate services designed to maintain community involvement and living, to the greatest extent possible. Such efforts can be done by the relevant state agencies alone, or in concert with the state's unit on aging, provider organizations, and the state developmental disabilities planning council.

To this end, the National Task Group recommends that states:

• Develop strategic plans for addressing the increasing rate of dementia among state residents with an intellectual disability.

• Encourage development of community-based care settings that are specifically designed for dementia care as well as those settings that are supportive of persons with dementia and allow for aging in place.

• Establish shared or regional assessment and diagnostic resources to help provide more accurate clinical information of affected persons.

5.0 Education and Training

The National Task Group recognizes the need for more information related to age-associated cognitive decline and neuropathologies (such as dementia), particularly how they apply to people with an intellectual disability and impact their families, friends, advocates and caregivers. Advocacy groups are the first to acknowledge that there is insufficient information available to caregivers, staff, and administrators related to this area. National organizations representing the interests of families and people with an intellectual disability have yet to fully acknowledge this emerging service need area and the lack of comprehensive consumer information. This dearth is troubling as it serves as a barrier to a better understanding of dementia and how it affects people with an intellectual disability. It does not allow for the provision of information that may include practical approaches to care planning or possible options for helpful service delivery.

The National Task Group also recognizes that there are too few trained personnel within the intellectual disabilities arena who are capable and comfortable with diagnosing, advising on, and helping with interventions and supports related to nuances of dementia and intellectual disabilities. The National Task Group views staff training and support as a critical issue in supporting persons with an intellectual disability affected by dementia in community settings. Organizations need to acknowledge a significant shift in skill set requirements and a change in philosophy of care when dementia-related behaviors predominate (as opposed to care philosophies inherent to lifelong intellectual disabilities). The National Task Group also recognizes that the knowledge related to, and the complications of, aging have not been cultivated or well developed in personnel who work within the intellectual disabilities arena. Greater emphasis needs to be put on training that will allow for early recognition and appropriate interventions and supports related to dementia and intellectual disabilities.

> **'It is what it is'... One physician's dilemma**
>
> *"I am a primary care physician who has been treating patients for 15 years. I do see a number of people with intellectual and developmental disabilities and have watched them age and do see that they are losing function over time. I am often asked whether they are becoming demented. I try to compare them to my other patients, but it is extremely challenging to diagnose and treat without any guidelines or definitive objective outcome measures. I would like to work with the various agencies but I don't have a relationship with them. I would like to refer to developmental neurologists for advice but there are none in my area. I feel uncomfortable with this complex patient population; the information I get is vague and the patients most often can't explain what bothers them. I would like to get more information on guidelines and care practice standards so that I can do a better job in making a difference in their lives."*

The National Task Group firmly believes that various professionals and organizations can help with training and disseminating information. It proposes that state developmental disabilities agencies and university centers on disabilities, as well as national advocacy associations, collaborate on a national initiative to produce consumer based information, enhance the training of professionals, and create regional consultative resources that can provide more accurate diagnoses and program planning supports. A national training curriculum and education initiatives would go far to meet this end. Also, a national effort to improve the competency of agencies and staff in developing and delivering 'dementia-capable' services is needed.

Such initiatives would ensure that staff develop and have knowledge and skills related to aging as it affects adults with an intellectual disability, and have a good grounding in dementia and dementia-capable care. This should include a basic understanding of normal and atypical aging process from a biological, social, and psychological perspective; a basic understanding of cognitive decline (including early onset features), mild cognitive impairment, and dementia in its different forms, along with their different manifestations and progressions; an understanding of best practices for providing day-to-day specialized dementia support; a familiarity with adaptations and modifications to the physical and social environments as well as to activities so as to promote active engagement; and a familiarity with the range of resources available to help support people with dementia and their families.

The National Task Group proposes that the nation's network of the University Centers on Excellence on Disabilities (UCEDs) and Leadership in Neurodevelopmental and Related Disabilities (LEND) programs can help by instituting more training within their state for providers, clinical personnel, families, and administrators as to the issues related to dementia, care practices, and medical knowledge of dementia and its associated and underlying causes. Similarly, the National Task Group would expect that the myriad Centers on Aging within the nation's universities would also help promote community education and a better understanding of the aging-related problems facing adults with an intellectual disability.

It is proposed that allied groups collaborate and coordinate efforts to promote training and education, as well as produce informational materials that could be disseminated in print as well as being web-based. The National Task Group would encourage national groups such as the AADMD, the AAIDD, the DDNA, and others to develop training workshops, seminars and webinars to help in exposing families, professionals, and staff to a broad base of knowledge on this topic.

Recommendation #18: Develop a universal curriculum, applicable nationwide, on dementia and intellectual disabilities geared toward direct care staff, families, and other primary workers.

Recommendation #19: Organize and deliver a national program of education and training using workshops and webinars, as well as other means, for staff and families.

Recommendation #20: Develop and produce an education and information package for adults with an intellectual disability to help them better understand dementia

6.0 Financing

Alzheimer's disease is a costly condition. The Alzheimer's Association estimates that the annual cost of care in the United States is $183 billion, with an additional amount attributed to the unreimbursed expenses incurred by families. The National Task Group recognizes that some portion of these costs are assumed by families caring at home for adults with an intellectual disability and the balance by public and private entities providing for some form of supports or care. Of the nearly 15 million Americans providing unpaid care for a person with dementia, about a quarter of one percent are parents or other family members of an adult with an intellectual disability. Many of these families have been providing life-time care and supports, most without any substantial financial assistance from public resources.

Two main resources exist to enable families to maintain the care and supports they are providing at home, as well as supporting similar efforts of provider agencies. The resources are funds provided by the Home and Community Based Waiver services via the Medicaid program and state appropriations to state developmental disabilities authorities. Within the current fiscal climate in the nation, both at the federal and state level, both of these resources will be challenged to expand their support of vulnerable adults, in particular those with an intellectual disability and dementia. The National Caregiver Support Program is also available, but offers limited support services directed at the caregivers.

The financial challenges will include adjusting reimbursement rates to intellectual disabilities' providers having to provide immediate and long-term care to adults with dementia. The reality is that with progressive decline and aggravated behavioral demands, the daily cost of care will increase beyond present reimbursement rates – at least for those adults with mid-state dementia (the stage with the most behavioral management demands on staff). However, states will need to adapt their costing mechanisms and adjust their reimbursement rates accordingly as more provider agencies take on the responsibility of primary long-term care of adults affected by dementia. As noted by the administrator in the vignette, the ability to maintain individuals as they decline is in jeopardy due to funding cuts and the on-going economic challenge facing many localities

States, too, need to recognize the needs of many families who are still providing care at home and who are going to be challenged by their own aging, and the presence of dementia in their family member with an intellectual disability. An equitable adjustment of costs to help keep these families as viable caregivers will need to be made. The reality is that many caregivers' physical and emotional health as well as family finances can often be compromised and stretched beyond capacities to continue to cope and care at home for someone affected by dementia.

> **We were told by an agency administrator that cut-backs in financing are a threat to continued quality of care**
>
> *"At our agency we are firmly committed to an 'aging-in-place' philosophy, but it is enormously costly and resource intensive – as it often requires more demanding staffing patterns and major adaptations to the physical space. In our state, as persons with developmental disabilities can no longer be referred to skilled nursing facilities without a court order (as the result of a class action suit), 'aging-in-place' is de facto mandated. Unfortunately, money to pay for these services is not always available. At our agency, we have opted to make resources available to help our folks remain in their homes with the hope of eventually being able to recoup some of the monies – because we feel it is in their best interests as well as their housemates. We have found that while it can be enormously stressful on housemates to watch their friends decline, it is more disruptive to have the person move out. So, we have put more counseling supports in place for housemates and staff to help them manage the stress of living with or caring for a person with dementia. Thus far, we have been able to manage; however, we have had seven new cases of dementia diagnosed in the past three months and four of those folks live in homes that are not fully barrier-free. Our ability to maintain our 'aging-in-place' philosophy will be seriously challenged over the coming months; much will depend upon how quickly they decline. We are also looking at significant funding cuts in the coming year due to the ongoing economic woes in the state."*

With respect to the supports provided via the use of Medicaid funds, there is a disconnect between funding/regulation and care needs, leading to difficulty in adjusting both the types and the level of care provided and supporting changed staffing. What is needed is assurance that Medicaid funds assigned to persons with an intellectual disability will be acceptable for paying for the care and supports needed when those persons are recognized as also being affected and debilitated by dementia.

7.0 Possible Solutions

The National Task Group was created to both set in place a plan for addressing the growing challenge of helping people with an intellectual disability affected by dementia, and relating to the national agenda to be undertaken under the National Alzheimer's Project Act (NAPA). This document has laid out a series of issues and chartered a course of action that could be undertaken by many constituent bodies, both at the federal and state levels, to address the needs of people with an intellectual disability and dementia (or those at-risk). It is hoped that the federal Advisory Council on Alzheimer's Research, Care, and Services and the federal bureaucracy set up to enact the requirements of the NAPA will heed what the National Task Group has developed and recommended and will include the issues addressed by this report within the NAPA process.

The National Task Group is also cognizant that many national associations and organizations have a stake in seeing more constructive and useful services developed and provided to adults with an intellectual disability and dementia, their families, and caregivers, as well as the organizations trying to cope with an increasing number of persons affected. The National Task Group recognizes that the main organization chartered to advocate and provide information about Alzheimer's disease and related dementias, the Alzheimer's Association, is a key partner in this effort. Many local chapters have responded to the challenge and have developed special initiatives to aid families who are wrestling with care-at-home challenges when a relative has an intellectual disability and dementia – however, many of their local chapters have yet to undertake such efforts.

The National Task Group is aware that many parent-based organizations share the hope that the national planning and services provision process, under NAPA, will consider the needs of this group. Among these organizations are The Arc, the National Down Syndrome Society, and others. It is anticipated that these organizations will join in on the conversation related to this issue and add their weight to the consideration of the issue within the NAPA process.

It is also anticipated that other national associations, such as those representing the state agencies for developmental disabilities, as well as providers, and certification organizations will also recognize these issues as important and contribute their collective weight to this process.

The National Task Group proposes that state developmental disabilities councils can help by targeting this older age issue in their states, by highlighting the concerns of providers and families, by lobbying the state developmental disabilities agency, and by funding and initiating pilot projects associated with dementia. The National Task Group also proposes that national groups such as The Arc, the National Down Syndrome Congress, the National Down Syndrome Society and others adopt policies which would enhance and enable families to be better informed about dementia and its effects and life course in adults with an intellectual disability. It will be most helpful if these organizations undertake initiatives to (a) link with local Alzheimer's Association chapters in helping family caregivers, (b) produce and disseminate informational materials for families and other caregivers, and (c) support the development of community-based services that enhance remaining in the community and living with dignity.

8.0 National Dementia and Intellectual Disabilities Action Plan

The main aim of the National Task Group is to develop a roadmap and strategies of support and care that ultimately enhance the quality of life of adults with an intellectual disability affected by dementia. The subsidiary aims are to (a) reduce the adverse impact of dementia in the lives of adults with an intellectual disability and their caregivers, (b) enhance the quality of living environments where people with an intellectual disability and dementia reside, (c) educate the population at large about the lives of adults with an intellectual disability affected by dementia, and (d) increase the competency of the workforce with respect to dementia and dementia care.

To achieve these aims the National Task Group proposes the following action steps that should be undertaken to address the growing challenge of aiding and supporting adults with an intellectual disability affected by dementia, their families and caregivers, and the network of agencies that serve people with intellectual and developmental disabilities.

- At the federal level
    - ♦ The needs of adults with an intellectual disability and their caregivers be given consideration within the national plans and recommendations to Congress made by the federal Advisory Council on Alzheimer's Research, Care, and Services.

    - ♦ The Administration on Developmental Disabilities and the Administration on Aging agree on a consistent policy towards meeting the needs of adults with an intellectual disability and that they request their constituents consider the needs of adults with an intellectual disability in their respective state plans as required under the Developmental Disabilities Bill of Rights Act and the Older Americans Act.

- At the national level
    - ♦ Professional organizations undertake initiatives to produce and provide training and education, produce technical materials, and publish practice guidelines related to adults with an intellectual disability.

    - ♦ National advocacy associations undertake initiatives that promote a greater understanding of the stages of dementia, services most useful to adults with an intellectual disability, and develop a range of programs and supports to caregivers and staff.

- At the state level
    - ♦ State developmental disabilities authorities give due consideration to planning for, regulating, and financing specialized services for adults with an intellectual disability affected by dementia.

    - ♦ Developmental disabilities planning councils undertake initiatives that promote a greater understanding of programs and services aiding adults with an intellectual disability and their caregivers.

    - ♦ Area agencies on aging (AAA's) integrate the needs of adults with an intellectual disability and their family members into their policy, planning and program development processes.

- At the local level
    - ♦ Provider organizations and agencies undertake screening and assessment functions, as well as develop specialized services to aid adults with an intellectual disability and their caregivers.

With respect to the National Dementia and Intellectual Disabilities Action Plan, the National Task Group has formulated the following recommendations previously noted in this document:

A. **To better understand dementia and how it affects adults with an intellectual disability and their caregivers:**

> *Recommendation #1: Conduct nationwide epidemiologic studies or surveys of adults with intellectual disabilities that establish the prevalence and incidence of mild cognitive impairment and dementia.*
> *Who could do it: Federal agencies (Administration on Developmental Disabilities, Administration on Aging, National Institute on Disability and Rehabilitation Research)*
>
> *Recommendation #2: Conduct studies to identify and scientifically establish the risk factors associated with the occurrence of dementia among adults with an intellectual disability.*
> *Who could do it: Universities' academic and research centers*
>
> *Recommendation #9: Conduct studies on the impact of aging of family caregivers on the support and care of adults with intellectual disabilities residing in at-home settings.*
> *Who could do it: Universities' academic and research centers*
>
> *Recommendation #11: Conduct nationwide medico-economic studies on the financial impact of dementia among people with intellectual disabilities in various service provision settings.*
> *Who could do it: Universities' academic and research centers*

B. **To institute effective screening and assessment of adults with an intellectual disability at-risk, or showing the early effects of, dementia**

> *Recommendation #3: Develop guidelines and instructional packages for use by families and caregivers in periodically screening for signs and symptoms of dementia.*
> *Who could do it: American Academy of Developmental Medicine and Dentistry*
>
> *Recommendation #4: Encourage provider agencies in the United States to implement screenings of their older-age clientele with an intellectual disability who are at-risk of or affected by dementia.*
> *Who could do it: State developmental disabilities planning councils, State developmental disabilities authorities*
>
> *Recommendation #5: Examine the utility of adopting an instrument such as an adapted Dementia Screening Questionnaire for Individuals with Intellectual Disabilities for use annually in preparation for the annual wellness visit.*
> *Who could do it: Universities, Providers, American Academy of Developmental Medicine and Dentistry*

Recommendation #6: Conduct an evaluation of a workable scoring scheme for the Dementia Screening Questionnaire for Individuals with Intellectual Disabilities that would help identify individuals in decline.
Who could do it: Universities' academic and research centers

Recommendation #8: Promote the exchange of information among clinicians regarding technical aspects of existing assessment and diagnostic instruments for confirming presence of dementia in persons with an intellectual disability.
Who could do it: American Academy of Developmental Medicine and Dentistry, American Association on Intellectual and Developmental Disabilities, Developmental Disabilities Nurses Association

C. **To promote health and function among adults with an intellectual disability**

Recommendation #15: Develop and disseminate a set of nutritional and dietary guidelines appropriate for persons with an intellectual disability affected by dementia.
Who could do it: American Academy of Developmental Medicine and Dentistry

Recommendation #16: Develop and disseminate health practice guidelines to aid primary care physicians and related health practitioners address assessment and follow-up treatment of adults with an intellectual disability presenting with symptoms of dementia.
Who could do it: American Academy of Developmental Medicine and Dentistry, Developmental Disabilities Nurses Association

Recommendation #17: Conduct studies on the nature and extent of health compromises, conditions, and diseases found among adults with an intellectual disability and affected by dementia.
Who could do it: Universities' academic and research centers

D. **To produce appropriate community and social supports and care for adults with an intellectual disability affected by dementia**

Recommendation #10: Enhance family support services to include efforts to help caregivers to identify and receive assistance for aiding adults with an intellectual disability affected by dementia.
Who could do it: State developmental disabilities authorities, State units on aging, Area agencies on aging (AAAs), The Arc, National Down Syndrome Society

Recommendation #12: Plan for and develop more specialized group homes for dementia care as well as develop support capacities for helping adults affected by dementia still living on their own or with their family.
Who could do it: State developmental disabilities authorities

Recommendation #13: Plan and develop community-based dementia-capable supports to address the needs of those persons at-risk or affected by dementia.
Who could do it: State developmental disabilities authorities

Recommendation #14: Develop and disseminate social care practice guidelines to community agencies and professionals that address assessment, service development and life planning for adults with an intellectual disability presenting with symptoms of dementia.
Who could do it: American Association on Intellectual and Developmental Disabilities

E. **To produce a capable workforce and produce education and training materials**

Recommendation #7: Establish undergraduate, graduate, and continuing education programs, using various modalities, to enhance the diagnostic skills of community practitioners.
Who could do it: American Academy of Developmental Medicine and Dentistry, American Association on Intellectual and Developmental Disabilities, Council of Deans of Medical Schools and Allied Health Colleges

Recommendation #18: Develop a universal curriculum, applicable nationwide, on dementia and intellectual disabilities geared toward direct care staff, families, and other primary workers.
Who could do it: Administration on Developmental Disabilities, Universities, Developmental Disabilities Nurses Association

Recommendation #19: Organize and deliver a national program of training using workshops and webinars, as well as other means, for staff and families.
Who could do it: American Academy of Developmental Medicine and Dentistry, American Association on Intellectual and Developmental Disabilities, Developmental Disabilities Nurses Association, Universities

Recommendation #20: Develop and produce an education and information package for adults with an intellectual disability to help them better understand dementia.
Who could do it: American Academy of Developmental Medicine and Dentistry, Developmental Disabilities Nurses Association, Universities' academic and research centers

♥♥♥

9.0 Appendices

9.1 Notes

9.1.1 The National Task Group acknowledges the significant contribution of the following members in developing the Group's summative report: Kathy Bishop, Ph.D., Melissa DiSipio, MSA, Lucy Esralew, Ph.D., Lawrence T. Force, Ph.D., Mary Hogan, MAT, Matthew P. Janicki, Ph.D., Nancy Jokinen, Ph.D., Seth M. Keller, M.D., Ronald Lucchino, Ph.D., Philip McCallion, Ph.D., Julie A. Moran, D.O., Leone Murphy, MSN, Linda Nelson, Ph.D., Dawna T. Mughal, Ph.D., Nahib Ramadan, M.D., Kathy Service, Ph.D., Baldev K. Singh, M.D.

9.1.2 All names listed in the vignettes have been changed to protect the confidentiality of the persons and situations described. The National Task Group expresses its appreciation to the members who provided the vignettes used in this report and to the many families who proffered personal photographs of their relatives to use on the dedication page.

9.2 Listing of members of the National Task Group

James Acquilano, PsyD
NYS OWPDD
Finger Lakes DDSO
Rochester, NY

Tracy Aldridge, MD
Div. of Develop. Disabilities
Illinois Dept of Human Services
Springfield, IL

Edward F. Ansello, PhD
Virginia Commonwealth Univ.
Richmond, VA

Jill Baker
Cottonwood, Inc.
Lawrence, KS

Kathleen Bishop, PhD
University of Rochester
Rochester, NY

Kelly Bohlander, MSW, MBA
Pyramid, Inc.
Tallahassee, FL

Betty Boyko, RN, BSN
Fraser LTD
Fargo, ND

Ernest Brown, MD
Individual Development Inc.
Washington, DC

Thomas J. Buckley, EdD
Lucanus
Hollywood, FL

Douglas Buglewicz, MEd
North Mississippi
 Regional Center
Oxford, MS

Casey Burke
Seguin Services Inc.
Cicero, IL

Diana Burt, PhD
Consultant
Madison, WI

Nicole Cadovius, MBA
Ability Beyond Disability
Sherman, CT

Ann Cameron Caldwell, PhD
The Arc
Silver Spring, MD

Barbara Caparulo, PsyD
Institute of Professional Practice
Fitchburg, MA

Thomas Cheetham, MD
Surrey Place Center
Scarborough, ON

Vincent Chesney
Selinsgrove State Center
Selinsgrove, PA

Brian Chicoine, MD
Lutheran General Hospital
Park Ridge, IL

Larry Clausen, MA
Arizona Developmental
 Disabilities Planning Council
Phoenix, AZ

Cyndy Cordell, MBA
Alzheimer's Association
Chicago, IL

Paul Cotton, PhD
Paul Carey University
Hattiesburg, MS

Melissa Disipio, MSA
Philadelphia Coordinated
 Health Care
Philadelphia, PA

Lucille Esralew, PhD

Trinitas Regional Medical Center
Cranford, NJ
Kate Fialkowski
Administration on
 Developmental Disabilities
Washington, DC

Lawrence Force, PhD
Center on Aging and Policy
Mount Saint Mary College
Newburgh, NY

David Fray, MBA, DDS
Developmental Disabilities Div
Hawaii Department of Health
Honolulu, HI

Sue Gant, PhD
Gant Yackel and Associates
Hawarden, IA

Bill Gaventa, MAT
Boggs Center on
 Developmental Disabilities
New Brunswick, NJ

Rick Glaesser, MSW
University of South Florida
Tampa, FL

Andrew Griffin, PhD
Mexia State Supported Living
Center
Mexia, TX

Joan Earle Hahn, PhD.
University of New Hampshire
Durham, NH

Tamar Heller, PhD
University of Illinois at Chicago
Chicago, IL

E Adel Herge, OTD OTR/L
Thomas Jefferson University
Philadelphia, PA

Mary Hogan, MAT

Family Advocate
Eliot, ME

Matthew Janicki, PhD
University of Illinois at Chicago
Chicago, IL

Allison Jay, M.A., BCBA
Clinical Geropsychology
University of Colorado
Colorado Springs, CO

George Jesien, PhD
AUCD
Silver Spring, MD

Graeme Johnson, MB ChB
University of South Carolina
Dept. of Pediatrics
Columbia, SC

Nancy Jokinen, MSW, PhD
University of Northern
 British Columbia
Prince George, BC

Ted Kastner, MD
Developmental Disabilities
 Health Alliance
Bloomfield, NJ

Seth Keller, MD
AADMD
Cherry Hill, NJ

Angela King
Volunteers of America
Arlington, TX

Martha Fenn King, RN, CDDN
One Sky Community Services
Portsmouth, NH

Penny M. Kollmeyer, MPA
Consultant
Farmington, CT

Stephanie Kohl, CTRS

Mount Olivet Rolling Acres
Chanhassen, MN

Jim Kuemmerle, ACSW
Armstrong-Indiana MH/MR/EI
 Program
Kittanning, PA

Paul Landers
Pathfinder Village Inc.
Edmeston, NY

Frode Kibsgaard Larsen
Aging and Health, Norwegian
 Centre for Research, Education
 and Service Development
Kongsberg, Norway

Gwen Lee
RMS, Inc.
Cincinnati, OH

Ronald Lucchino, PhD
Utica College
Longboat Key, FL

Marian Maaskant, PhD
Pergamijn Foundation
Echt, the Netherlands

Cheryl Marsh
Sydney Creek Alzheimer's &
 Dementia Community Care
San Luis Obispo, CA

Philip McCallion, PhD
University at Albany
Albany, NY

Mary McCarron, PhD
Trinity College School of
 Nursing & Midwifery
Dublin, Ireland

Dennis McGuire, PhD
Lutheran General Hospital
Park Ridge, IL

Dawn McKenna
Down Syndrome Research
 Foundation
Burnaby, BC

Ana Christina Minerly, PhD
AHRC New York City
New York, NY

Julie A. Moran, DO
BIDMC Senior Health
Boston, MA

Tammie Morley, LCSW-R
Schenectady ARC
Schenectady, NY

Charles Moseley, EdD
NASDDDS
Alexandria, VA

Jan Moss
Center for Learning and
Leadership/UCEDD
Oklahoma City, OK
Dawna Torres Mughal, PhD
Gannon University
Erie, PA

Leone Murphy, MSN
Family Advocate
Stuart, FL

Linda Nelson, PhD, ABPN
University of California
 at Los Angeles
Los Angeles, CA

Maggie Nygren, EdD
AAIDD
Washington, DC

Isabelle O'Donoghue, MAppSci
Trinity College
Dublin, Ireland

Rita Ozbun
Developmental Homes
Grand Forks, ND

Paul Partridge, PhD
NYS OWPDD
Capital District DDSO
Albany, NY

Paul Patti, MA
NYS Institute for Basic Research
in Developmental Disabilities
Staten Island, NY

Kathryn G Pears, MPPM
Dementia-care Trainer
Phippsburg, ME

Elizabeth Perkins, PhD, RNMH
University of South Florida
Tampa, FL

Vee Prasher,,MD, PhD
University of Birmingham
Birmingham, United Kingdom

Paula Pratt
Cottonwood, Inc.
Lawrence, KS

Genny Pugh, MA, LPA
Turning Point Services
Morganton, NC

Kent Questad, Ph.D.
University of Washington
Shoreline, WA

Rick Rader
Kent Habilitation Center
Chattanooga, TN

Nabih Ramadan, MD
Nebraska Division of
 Developmental Disabilities

Beatrice, NE

Allen Ray
SimplyHome, LLC CMI, Inc.
Asheville, NC

Michael Rohr, EdD
CMI, Inc.
Somerville, TN

Stephanie Rosati-Pratico
Children's Hospital of
Philadelphia
Hamilton, NJ

Lou Ellen Ruocco, RN
Private Practice
Frontenac, MO

Catherine Rush, BS, MBM
Cuyahoga Co. Board of DD
Cleveland, OH

Rachael Sarto
University of Minnesota
Minneapolis, MN

Kathryn Service, RN, PhD(c)
Nurse Practitioner
Northampton, MA

Patricia Seybold
Kentucky Council on
 Developmental Disabilities
Frankfort, KY

Nancy Shanley
Ascend Management
Innovations
Nashville, TN

Baldev Singh, MD
Westchester Institute for
 Human Development
Valhalla, NY

Deborah Slater, MS
AOTA
Bethesda, MD

Julie Snyder
Tierra del Sol Foundation
Sunland, CA

Michael Splaine, MA
Policy Advisor
Alzheimer's Disease
 International
London, United Kingdom

Kathleen M Srsic-Stoehr, MSN
Family/Individual Advocate
McLean, VA

Claudia Stanley, RN, CDDN
Granite Bay Care
Portland, ME

Nancy Thaler
NASDDDS
Alexandria, VA

Lynne Tomasa, PhD, MSW
University of Arizona
Tucson, AZ

Leslie Udell
Winnserv, Inc.
Winnipeg, MB

Linda Ulinski, RN, CDDN
Philadelphia Coordinated
 Health Care
Philadelphia, PA

Renee Wachtel, MD, FAAP
Developmental Behavioral
 Pediatrics
San Leandro, CA

Kevin Walsh, PhD
Developmental Disabilities
 Health Alliance, Inc.
Vineland, NJ

Kara Walters
Cottonwood, Inc.
Lawrence, KS

Karen Watchman, PhD
University of Edinburgh
Edinburgh, United Kingdom

Fr. Dennis Weber
Catholic Social Services
Archdiocese of Philadelphia
Springfield, PA

Chris White
Road to Responsibility, Inc
Marshfield, MA

Mary Alice Willis
DDNA
Orlando, FL

Sara Wolfson, MSN
University of Nebraska
 Medical Center
Omaha, NE

Ric Zaharia, PhD
Private Practice
Tucson, AZ

Lou Zimmer
UCP of Nassau County
Roosevelt, NY

David Zucker, EdD
Private Practice
Cincinnati, OH

9.3 References

[1] This definition is adapted from one used in Operational *Definition Recommendations Report*, developed by University of Massachusetts Medical School and Human Services Research Institute as part of the 2010 Research Topic of Interest (RTOI): Health Surveillance of Adults with Intellectual Disability, awarded by the Association of University Centers on Disabilities (AUCD) and funded through a cooperative agreement with the Centers for Disease Control and Prevention (CDC) National Center on Birth Defects and Developmental Disabilities (NCBDDD).

[2] Zigman, W.B., Schupf, N., Devenny, D., et al. (2004). Incidence and prevalence of dementia in elderly adults with mental retardation without Down syndrome. *American Journal on Mental Retardation*, 109, 126-141.

[3] Janicki, M.P.& Dalton, A.J. (2000). Prevalence of dementia and impact on intellectual disability service. *Mental Retardation*, 38(3), 276-288.

[4] National Down Syndrome Society. http://www.ndss.org/index.php?option=com_content&view=article&id=180&showall=1

[5] Alvarez, N. (2011*). Alzheimer disease in Down syndrome*. Medscape Reference: Drugs, Diseases and Procedures. http://emedicine.medscape.com/article/1136117-overview#aw2aab6b4aa

[6] Braddock, D., Hemp, R., Rizzolo, M.C., Haffer, L., Tanis, E.S., & Wu, J. (2011). *The state of the states in developmental disabilities 2011*. Washington, DC: AAIDD.

[7] Alzheimer's Association . (2011). 2011 Alzheimer's disease facts and figures. *Alzheimer's & Dementia*, 7(2), 1-63. Chicago: Author.

[8] Janicki, M.P., & Dalton, A.J. (1999). *Aging, dementia, and intellectual disabilities*: A Handbook. Philadelphia: Taylor and Francis

[9] Wilkinson, H.A., & Janicki, M.P. (2002). The Edinburgh principles with accompanying guidelines and recommendations. *Journal of Intellectual Disability Research, 46*, 279-284.

[10] Deb S., Hare M., Prior L. & Bhaumik S. (2007). Dementia screening questionnaire for individuals with intellectual disabilities (DSQIID). *British Journal of Psychiatry*, 190, 440-444.

[11] PCAD project. (2012). Protocol for recording baseline behavior information for persons with Down syndrome. http://www.rrtcadd.org/resources/RRTCADD/2012-Protocol.pdf.

[12] Kinsky-McHale, S., Urv, T., Zigman, W., & Silverman, W. (2010, August). Neurological symptoms associated with mild cognitive impairment in adults with Down syndrome. Paper presented as part of the symposium, *Transitions in neurological and cognitive phenotypes and dementia status in people with Down syndrome*, at the 118th annual convention of the American Psychological Association, San Diego, California, August 14, 2010.

[13] Prasher, V.P. (2005). *Alzheimer's disease and dementia in Down syndrome and intellectual disabilities*. Milton Keynes: Radcliffe.

[14] Mosely, C. National core indicators: A review of the data on persons with Alzheimer's syndrome. Presentation at the National Task Group meeting on June 6th, St. Paul, Minn.

[15] US Surgeon General. (2002). *Closing the gap: A national blueprint to improve the health of persons with mental retardation*. Rockville, MD : U.S. Department of Health and Human Services.

[16] Krahn, G. L., Hammond,L., & Turne, A. (2006). A cascade of disparities: Health and health care access for people with intellectual disabilities.*ental Retardation and Developmental Disabilities Research Reviews*. 12(1) 70-82.

[17] LHIDS study reports. University of Chicago. http://www.rrtcadd.org/Research/Health_Function/Cohort/LHIDS/Info.html

[18] Janicki, M.P., Heller, T., Seltzer, G.B., & Hogg, J. (1996). Practice guidelines for the clinical assessment and care management of Alzheimer's disease and other dementias among adults with intellectual disability. *Journal of Intellectual Disability Research*, 40(4), 374–382.

[19] Group home care for adults with intellectual disabilities and Alzheimer's disease. *Dementia - The International Journal of Social Research and Practice*, 2005, 4(3), 361-385.

[20] Quality outcomes in group home dementia care for adults with intellectual disabilities. *Journal of Intellectual Disability Research*, 2010, 55(8), 763-776.

9.4 Abbreviations

AAA: Area agencies on aging
AAIDD: American Association on Intellectual and Developmental Disabilities
AADMD: American Academy of Developmental Medicine and Dentistry
DDNA: Developmental Disabilities Nurses Association
LEND: Leadership in Neurodevelopmental and Related Disabilities
MCI: Mild cognitive impairment
NAPA: National Alzheimer's Project Act
NTG: National Task Group
RRTC: Rehabilitation Research and Training Center
UCED: University Centers on Excellence on Disabilities

♥♥♥

www.ingramcontent.com/pod-product-compliance
Lightning Source LLC
Chambersburg PA
CBHW051101180526
45172CB00002B/735